A Pre-Engineering Guide to Pro/ENGINEER Wildfire 4.0

CURRICULUM & COURSE-BASED TEXTS & RESOURCES DIVISION

Alliance for Catholic Education Press
at the University of Notre Dame

A Pre-Engineering Guide to Pro/ENGINEER Wildfire 4.0

Avery J. Scott
Richard B. Strebinger

ALLIANCE FOR CATHOLIC EDUCATION PRESS
at the University of Notre Dame

Notre Dame, Indiana

Alliance for Catholic Education Press
at the University of Notre Dame
158 I.E.I. Building
Notre Dame, IN 46556
http://www.nd.edu/~acepress

ISBN: 978-0-9788793-9-6

Text design by Julie Wernick Dallavis
Cover design by Mary Jo Adams Kocovski

Library of Congress Cataloging-in-Publication Data

Scott, Avery J., 1987-
  A pre-engineering guide to Pro/ENGINEER Wildfire 4.0 / Avery J. Scott,
Richard B. Strebinger.
      p. cm.
  Summary: "Provides step-by-step lessons and instructions for high
school students using the computer aided design software, Pro/ENGI-
NEER Wildfire 4.0"--Provided by publisher.
  ISBN 978-0-9788793-9-6 (pbk. : alk. paper)
  1. Pro/ENGINEER. 2. Computer-aided design. 3. Mechanical drawing.
I. Strebinger, Richard B., 1958- II. Title.
  TA174.S395 2009
  620'.00420285536--dc22
                        2009014726

This book is printed on acid-free paper.

Printed in the United States of America.

Table of Contents

Introduction

This book was created to introduce high school students to the broad capabilities of modern computer aided design software. Specifically, these lessons act as a step-by-step tutorial that will establish a foundation in the use of Pro/ENGINEER Wildfire 4.0 developed by Parametric Technology Corporation (PTC).

As students explore the Wildfire environment, they will be actively engaged creating parts. After all of the individual parts have been created, they will be assembled into a remote control device shown below. Each new chapter builds on previous lessons to aid students in building a concrete foundation while working with advanced topics. A synthesis of instruction and hands-on experience is critical to developing the independence needed to take full advantage of this powerful software. This tutorial exposes students to something they can relate to and starts them down the path of modeling the objects surrounding them.

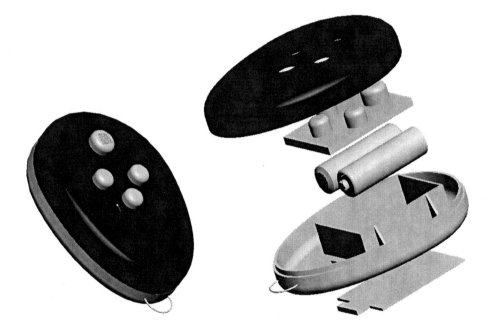

This guide becomes an excellent resource for students. After receiving in-class instruction, students often find themselves forgetting small but necessary details. With this book, students can turn through the various lessons and find step-by-step material to assist them in recreating multiple features.

Before beginning instruction, make sure that students have a consistent directory to store their files. If there is a concern for limited hard drive space, consider having students bring a flash drive to class each day. For assemblies to function properly in Pro/ENGINEER Wildfire 4.0, the assembly and all of its individual part files need to be stored in the same location.

Starting Pro/ENGINEER Wildfire 4.0

The following screen is presented upon opening Pro/ENGINEER Wildfire 4.0 (PE 4).

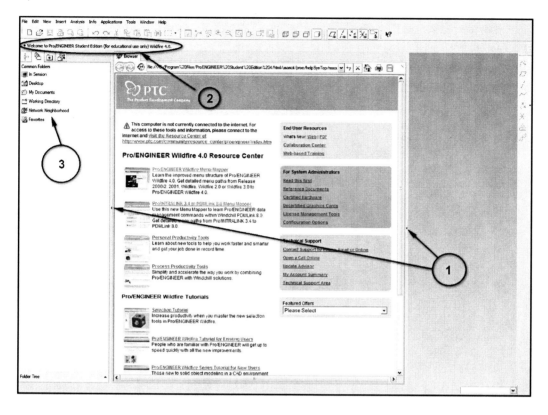

The main portion of the screen is occupied by the PE4 Internet browser. This acts just as any other browser. If the computer is connected to the Internet, it should default to PTC's resource center. If it fails to do so and you have a live Internet connection enter,

http://www.ptc.com/community/resource-center/proengineer/

into the Address box. The Internet browser is a useful utility when looking for ways to advance your PE4 skills.

The sets of arrows in the screen labeled "1" will minimize and maximize the windows on the screen. Clicking on the arrows pushes the window in the direction indicated. Minimize the Internet browser by left-clicking on the arrow set toward the right side of the image.

The area labeled "2" at the top of the image is the command prompter for Pro-Engineer. This will display feedback on most of the actions taken while using the software. When stuck on something, this can be very useful as it frequently prompts and guides the user to make specific selections and take other courses of action.

The window labeled "3" is the Folder Navigator. It allows you to look for files without minimizing the Pro-Engineer software. The Navigator aids the user in choosing a working directory. This window switches to the Model Tree when work begins on a part or assembly file which will soon become apparent.

Part 1—Battery Cover

First, the establishment of a **Working Directory** is needed. This will dictate the storage location of newly created files. Also, the Working Directory becomes the default folder when opening part files in PE4. Storing all part files in the same directory is essential for creating an assembly which will be Part 6 of this tutorial.

Locate **File** in the menu bar at the top left of the screen. Select **File** > **Set Working Directory** as shown in Figure 1-1.

Figure 1-1

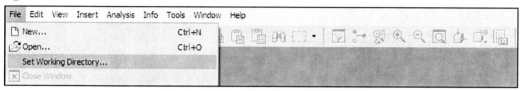

The **Select Working Directory** browser opens. The course instructor should indicate the directory to be used (the folder where class files should be stored). The icons at left shown in "1a" (see Figure 1-2) allow for quick browsing. When the proper folder has been found, open it and select **OK**. Notice that the prompter at the top of the screen indicates that the directory has been successfully changed.

Figure 1-2

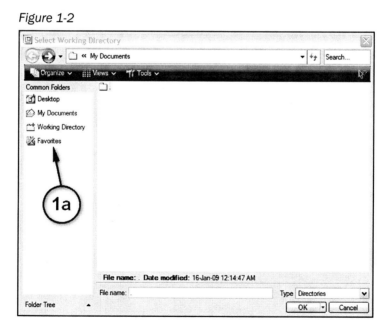

**NOTE: THE ABOVE STEP NEEDS TO BE REPEATED
EVERY TIME PE4 IS LAUNCHED.**

New parts, sketches, drawings, and assemblies are all initiated in the same way. A user can either select **File** > **New** (see Figure 1-3) or use the **Create a new object** shortcut circled in Figure 1-4.

Figure 1-3

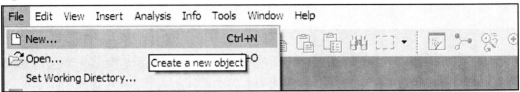

Figure 1-4

The **New** dialog box appears (see Figure 1-5). If it is not already set, be sure to select **Part** as the **Type**, **Solid** as the **Sub-Type** and **Use default template** as indicated below. In the **Name** text box enter "Battery_Cover" (no quotes). Notice the use of an underscore in the name of this part. PE4 will not accept names with spaces as well as certain characters. For example, the use of quotes, slashes, periods, and colons is prohibited. If an illegal character is used, PE4 will create a notification that a different name must be used. Select **OK** in the **New** dialog box to complete the initiation of the new part.

Figure 1-5

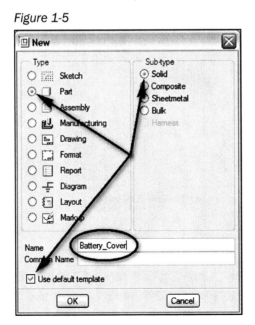

The PE4 3-D modeling environment is activated, represented by the space labeled "1b" (see Figure 1-6). PE4 allows the user to use this space to freely orient the part.

The jumble of lines in the center of the 3-D environment labeled by "2b" is the datum plane set. The three datum planes (each individually circled **FRONT**, **RIGHT**, and **TOP**) are all perpendicular to one another forming a 3-D space. They exist as a base of references around which each part is developed.

Figure 1-6

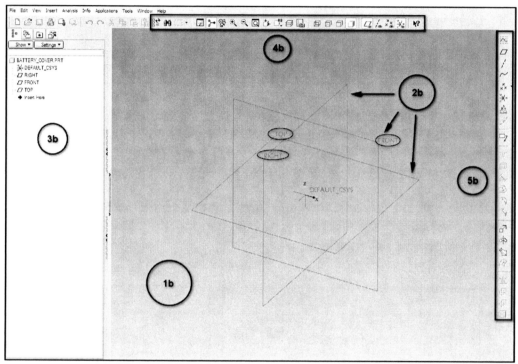

As mentioned earlier, the area now labeled "3b" has changed into the battery cover's model tree. Since nothing has been created yet, the model tree only consists of the basic default features; a central coordinate system and three datum planes.

There are two primary toolbars containing shortcuts. The horizontal toolbar at the top of the screen ("4b") contains options concerning the display of the model and its references (datum planes, axes, and coordinate systems) in the 3-D environment.

The vertical toolbar on the right side of the screen ("5b") contains shortcuts to create sketches, extrusions, and other more advanced features in PE4. The vertical toolbar will be used more frequently than the other.

A part can be thought of as a combination of extrusions with various magnitudes and directions. An extrusion, a 3-D feature, must be based on a 2-D sketch. For example, a cube is created by sketching a square and extruding it (extending the sketch from a 2-D to a 3-D figure). Likewise, a cylinder is created by sketching a circle and extruding that. Thus, you will begin to develop the battery cover by first creating a sketch and then extruding it.

Left-click the **Sketch Tool** located at the top of the vertical toolbar as shown in Figure 1-7.

Figure 1-7

The **Sketch** dialog box opens (see Figure 1-8). In order to enter the sketching environment, PE4 first requires a selection of the plane to be sketched on. Move the mouse over the **TOP** datum plane in the 3-D environment and left-click to make the selection. The name **TOP** should highlight in a light blue color.

Notice that the text box next to **Plane** (see Figure 1-9) indicates that you have selected the **TOP** plane. Select **Sketch** at the bottom of the **Sketch** dialog box to enter the sketching environment.

Figure 1-8   *Figure 1-9*

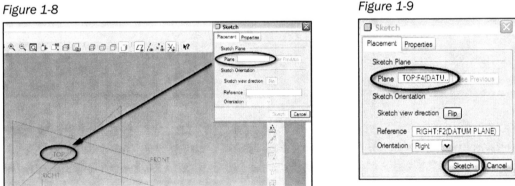

The environment orients so that you are looking at the **TOP** plane (1c). Notice that the **RIGHT** and **FRONT** datum planes (circled in Figure 1-10) are now represented by dashed lines. These lines are the base references around which the sketch will be built. These base references are necessary so that the location of the sketch in the 3-D environment is clearly defined.

First, a vertical centerline needs to be constructed along the **RIGHT** datum reference in the center of the window. This centerline allows for the automatic

Figure 1-10

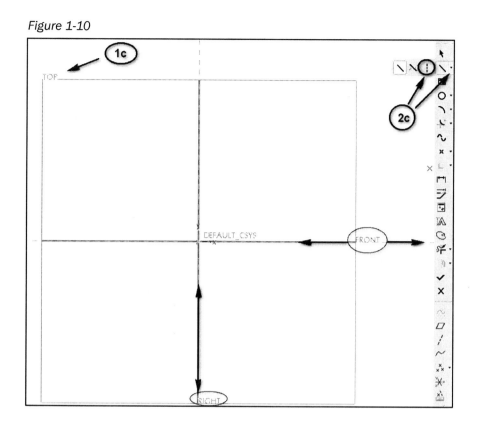

creation of symmetry constraints on the sketch as it is being drawn. A sketch's constraints combine with its dimensions to define it and make it unique. Most constraints are geometric. For example, the user can draw a circle and impose a geometric constraint that the circle be tangent to an existing line. This consequently will constrain the diameter of the circle since the center is fixed.

Left-click the small downward pointing arrow next to the line tool shown in Figure 1-10 in order to activate the tools menu tree. From the menu tree, select the dashed line (**Centerline** tool) indicated by "2c" in the image above.

In the sketching environment, move the mouse pointer on top of the vertical dashed line until a small red circle appears as shown in Figure 1-11. The red circle indicates that a constraint is being placed on the starting point of the line. This constraint specifically means that the point is on the **RIGHT** datum reference.

Left-click to begin creating the line. Notice that the line is horizontal. The line can be rotated by moving the pointer in a circular motion around the screen. Try moving the line into vertical position with this method. The centerline is vertical and in line with the **RIGHT** datum reference when the red bars appear on either side of the line as shown in Figure 1-12. The red bars represent another constraint indicating that the centerline is parallel to the **RIGHT** datum reference (in this case, vertical).

Figure 1-11

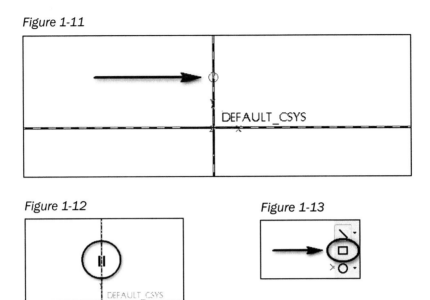

Figure 1-12   *Figure 1-13*

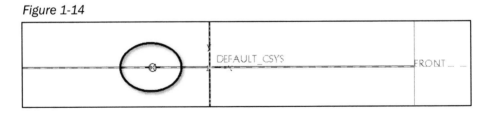

Left-click again to complete the line. Next select (left-click) the **Create rectangle** shortcut (see Figure 1-13) located directly below the line tool.

This time, in the sketching environment, move the pointer slightly to the left of the centerline and on the **FRONT** datum reference as shown in Figure 1-14.

Figure 1-14

Left-click to start creating the rectangle. Move the mouse pointer upward and to the right of the screen to expand the rectangle.

As the mouse is pushed to the right, a point will be reached which is the same horizontal distance from the centerline as the starting point. Two red arrows will appear on the top line of the rectangle to indicate a symmetry constraint about the centerline being placed on the rectangle. Stop expanding the rectangle once this constraint appears (see Figure 1-15).

Figure 1-15

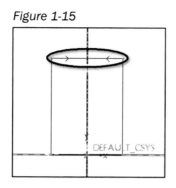

NOTE: Make sure there are not equal distance constraints on the horizontal and vertical lines. These will appear as an **L1** on both lines (see Figure 1-16). To get rid of this constraint, move the mouse pointer upwards until the figure is no longer a square. If you finish sketching a figure and realize that unwanted constraints have been formed, select them and hit delete on the keyboard.

Figure 1-16

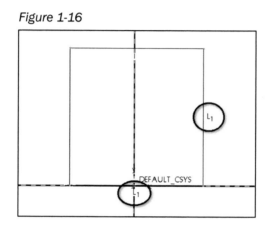

Left-click again to complete the rectangle.

Choose the **Select** (pointer) tool at the top of the vertical toolbar (see Figure 1-17).

Figure 1-17

Notice that dimensions now appear on the rectangle. Double-clicking a dimension allows the user to edit it by changing the value in the text box.

Change the horizontal dimension to a value of **2** (see Figure 1-18). Press Enter on the keyboard.

Figure 1-18

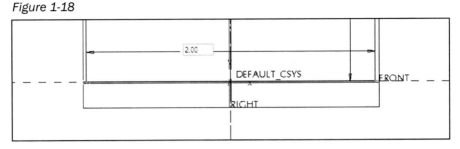

Depending on the value of the vertical dimension, the rectangle may seem to disappear. However, this appearance is simply the result of the vertical dimension being disproportionately larger than the horizontal dimension.

Change the vertical dimension to a value of **3** and press **Enter**. The rectangle should appear like Figure 1-19. You may need to zoom in (roll the mouse wheel backwards with the pointer centered over the image) to achieve this view.

Figure 1-19

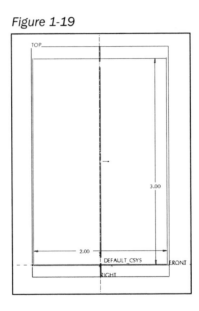

The rectangular sketch is now complete and ready to be extruded. To exit the sketching environment, select the **Check Mark** in the vertical toolbar in Figure 1-20.

Figure 1-20

PE4 returns to the 3-D modeling environment.

Locate and select the **Extrude** tool in the vertical toolbar shown in Figure 1-21.

Figure 1-21

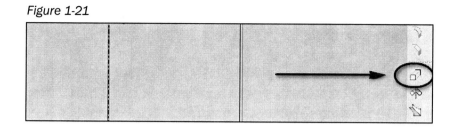

A new toolbar (Figure 1-22) appears in the top left corner of the screen.

Figure 1-22

The text box circled in Figure 1-22 dictates the depth of the extrusion. Click inside of the text box to edit the value to **.05**.

Select the **Green Check** at the top right (circled in Figure 1-22) to complete the extrusion.

To see the extrusion the part needs to be reoriented out of the current, flat view. Go to the menu at the top of the window and click on **View** to activate the drop-down menu. Highlight **Orientation** and choose **Standard Orientation**. Now it is easier to see the full geometry of the 2 x 3 x .05 solid panel (see Figure 1-23).

Figure 1-23

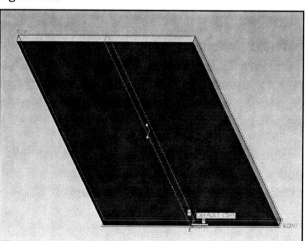

The model can be manipulated through the use of the middle mouse wheel.

Center the cursor over the middle of the part. Try rolling the middle mouse wheel forward and backward. Forward zooms out, backwards zooms in. This may seem counter-intuitive, but just relate it to pushing and pulling. Forward pushes it away, backward pulls it toward you.

Keep the cursor in the silver modeling area, but offset it to the upper-right, away from the part. Now slide the wheel forward and back. Notice how you can now push it into the upper-right corner, or pull it into the lower-left corner of the screen.

Pressing down on the mouse wheel and moving the mouse about the modeling view will spin the object. Between the spinning and zooming features, every aspect of the part can be seen.

To slide the part around the screen, hold the **Shift** key and **middle mouse button**, then move the mouse right or left. A red line will appear showing the trajectory of your drag.

If you lose track of the model, remember that you can always reset the view to the standard orientation as before or by using **Ctrl + D**. Also, you can activate the **Refit** icon in the horizontal toolbar as shown in Figure 1-24.

Figure 1-24

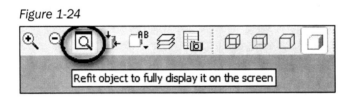

Notice how the model tree (shown in Figure 1-25) now contains a sketch (**Sketch 1**) and extrusion (**Extrude 1**). If you select the + next to **Extrude 1** you will see that **Sketch 1** appears under it. PE4 functions on **Parent/Child** relationships. Since the extrusion cannot exist without the sketch, the parent is the sketch and the child is the extrusion.

Suppose that the width of the battery cover needs to be changed without changing the depth. PE4 allows the user to edit **Sketch 1** and will automatically apply those changes to the extrusion, because of the **Parent/Child** relationship.

Select either of the objects labeled **Sketch 1**. Press the right mouse button to activate the drop down menu and select **Edit Definition** (see Figure 1-26).

Figure 1-25

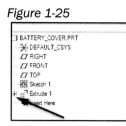

Figure 1-26

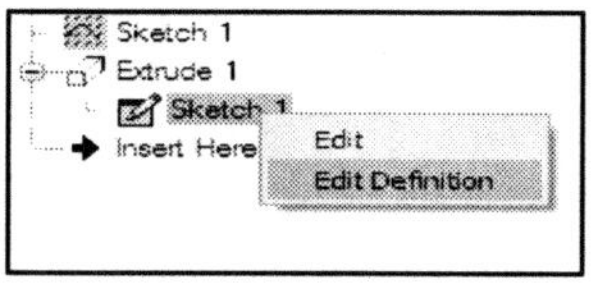

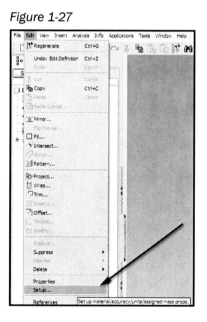

The sketching environment reopens. Double click the horizontal dimension to change it from **2** to **1.4**. Do the same for the vertical dimension, changing it from **3** to **2**. Select the **Check Mark** as before. If you have trouble, refer back to page 12. PE4 returns to the 3-D environment and the extrusion is regenerated accordingly.

It is important to know what units are used when assigning values to dimensions. PE4 allows the user to control the standard units for length, mass, and time.

In the menu bar at the top of the screen, select **Edit** > **Setup** (see Figure 1-27).

Figure 1-27

The **Menu Manager** pops up on the right side of the screen. In this box select **Units** (see Figure 1-28).

Figure 1-28

The **Units Manager** window displays all of the unit convention options. For the purposes of this project, the Pro/E Default units (**Inch lbm Second**) are desired. If your units are not already set (highlighted with a red arrow) appropriately, click the **Pro/E Default** units and choose **Set** (see Figure 1-29).

When the units are switched, the **Changing Model Units** dialog box opens. Be sure to select the **Interpret dimensions** option and then pick **OK** (see Figure 1-30). Select **Done** in the **Menu Manager**.

Figure 1-29

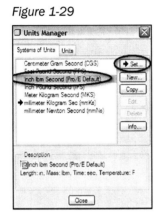

Figure 1-30

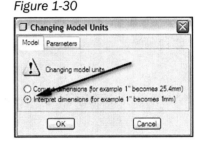

The battery cover is now **1.4 x 2 x .05** inches.

The final cover needs to have some additional features and look like the part shown in Figure 1-31.

Figure 1-31

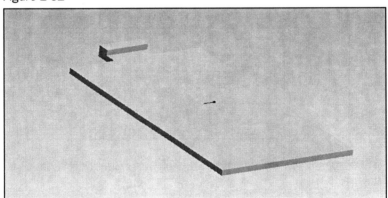

To accomplish this final form, you will need to edit the original sketch again as well as create another sketch-extrude pair.

At this point, the lesson will move faster since most of the following steps have already been covered once. If you have trouble, go back to the earlier pages and go through the steps in more detail.

Select **Sketch 1** in the model tree. Press the right mouse button and choose **Edit Definition** as before.

Start a new rectangle on the top line of the original sketch just left of the centerline. The starting point is on top of the line if the small red circle appears as before.

Draw a rectangle like the one in Figure 1-32. Make sure that the symmetry constraint is applied as before. Select the pointer tool and set the horizontal dimension to **.6** and the vertical dimension to **.3**.

Figure 1-32

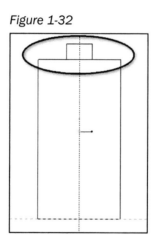

PE4 seldom allows the use of more than one closed figure in a sketch. In any case, two closed sketches cannot border one another. The common line between the two rectangles needs to be removed. For this a new, **Dynamic trim** tool is used. Selecting a line segment while this tool is active will cause the segment to be removed. The **Dynamic trim** tool is located in the vertical toolbar as shown in Figure 1-33.

Figure 1-33

After selecting the tool, move to the bottom line of the top rectangle that was just created. Left-click twice on each of the two segments in the image in Figure 1-34. Two clicks are needed to eliminate both of the overlapping segments.

Figure 1-34

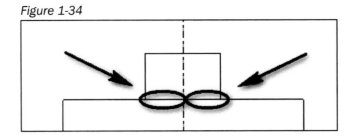

The result should be like Figure 1-35.

Figure 1-35

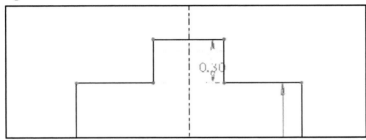

Note: Sometimes a sketch is accidentally spun into an unwanted position like the one shown in Figure 1-36. If this ever happens, select **View** > **Orientation** > **Sketch Orientation** *from the menu at the top of the sketch.*

Figure 1-36

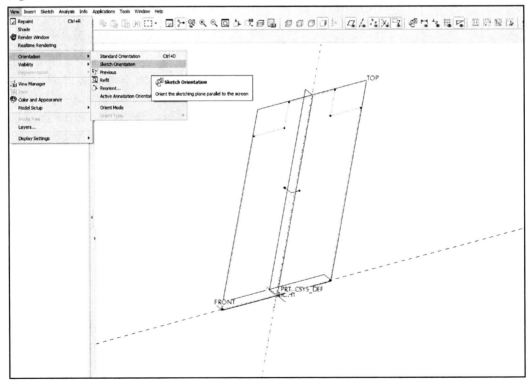

Select the check mark to complete the sketch. Again, notice that the **Parent/Child** relationship causes the extrusion to update.

Instead of sketching on the **TOP** datum plane as before, a new sketch will be created on the part surface. That way, the extrusion will start at the surface and move outward to give the additional depth to the last rectangle on the battery cover tab.

Locate the **Selection Filter** in the bottom right corner of the screen and activate the drop down menu. Choose **Geometry** from the menu (as shown in Figure 1-37).

Figure 1-37

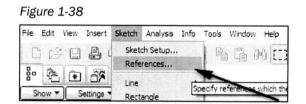

Left click on one of the two large surfaces (1.4 x 2) in the extrusion to make the selection. Notice that the surface changes color to pink to indicate the selected surface.

Select the **Sketch** tool and choose **Sketch** in the **Sketch** dialog box.

This new rectangle should be even with the top and both vertical sides of the small tab that was just created. To make this happen, create reference lines on the perimeter of the tab that start points can lock on to.

In the menu bar, select **Sketch** > **References** (see Figure 1-38).

Figure 1-38

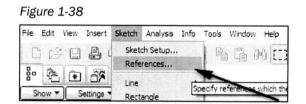

PE4 now requires the selection of edges for the formation of the reference lines. Select each of the edges indicated by the three arrows as shown in Figure 1-39. The edges will turn light blue when the pointer moves over them indicating that they are ready to be selected.

Figure 1-39

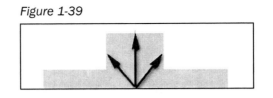

Clicking on each of these edges will yield three reference lines appearing as shown in Figure 1-40. **Close** the **References** dialog box.

Figure 1-40

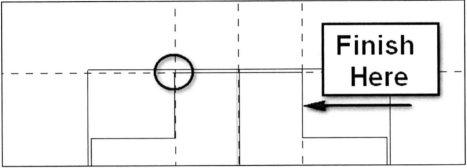

Start a new rectangular sketch at the intersection of the reference lines circled in Figure 1-40. Complete the rectangle about half way down the line on the right.

Select the pointer as before and edit the vertical dimension to **.15**. There is no need to edit the horizontal dimension since the figure is constrained by the reference lines.

Select the **Check Mark** to complete the sketch.

Select the **Extrude** tool. **Edit the** depth by double-clicking the value directly in the 3-D environment as if **editing** a sketch dimension (see Figure 1-41). Edit this value to **.05**.

Figure 1-41

If the extrusion is moving into the part, the direction can be edited in the toolbar in the upper left where the depth text box is located.

Try selecting the directional tool (Figure 1-42) and watch what happens to the preview of the extrusion.

Figure 1-42

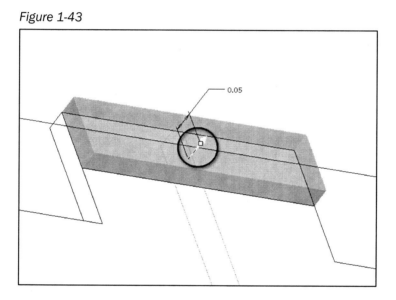

The same effect will occur if the yellow arrow in the 3-D environment (Figure 1-43) is clicked.

Figure 1-43

Choose the green **Check Mark** to complete the extrusion.

Select **File** > **Save** or hit the combination of keys **Ctrl + S**.

Notice in Figure 1-44 that the default file name in the **Save Object** dialog box is the part name (BATTERY_COVER). The ".prt" extension will be recognized by PE4 as well as other CAD software packages. Also note that the file will be stored to the working directory folder that you selected on the first page of this tutorial.

Figure 1-44

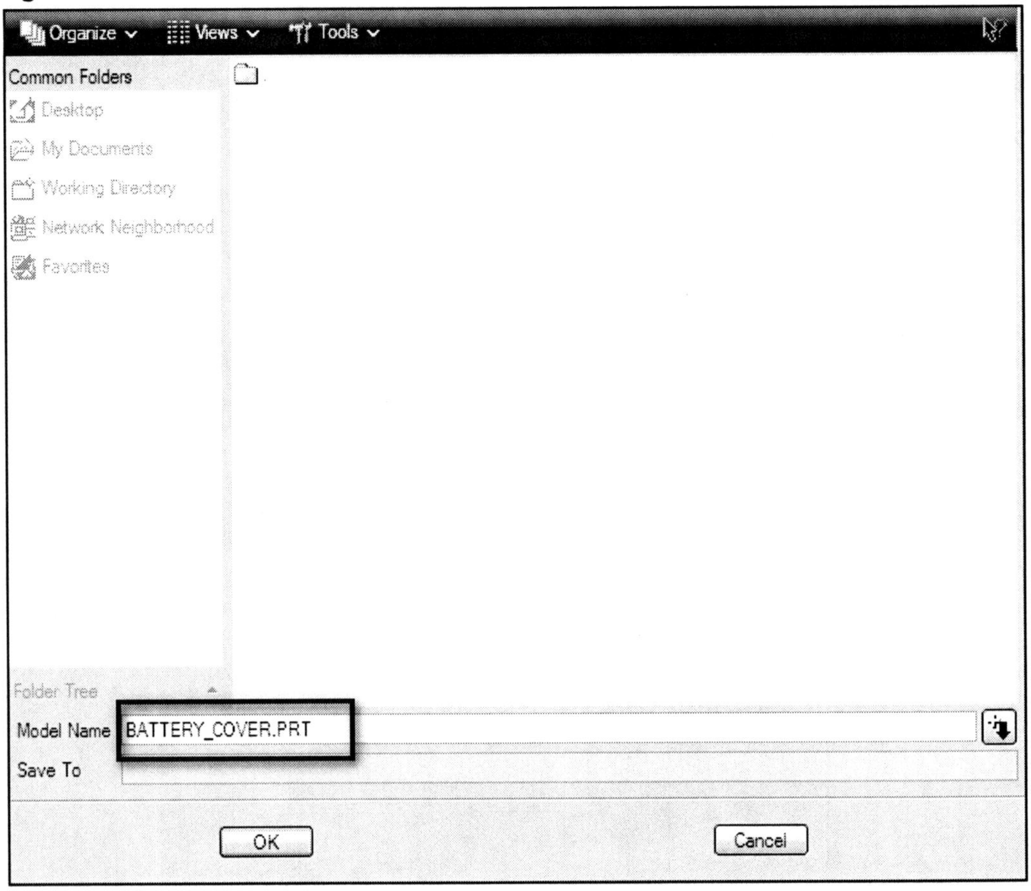

Choose **OK** to complete the save.

Congratulations! You have completed the first of five parts and are on your way toward finishing the assembly. Next you will create the batteries for the device. This exercise will reinforce your newly developed skills.

Part 2a—Battery (1)

If you are continuing Part 2 directly from Part 1 (without closing PE4), follow the "Option 1" instructions. Otherwise follow "Option 2."

Option 1:

Select **File** > **Close Window** as shown in Figure 2-1.

Figure 2-1

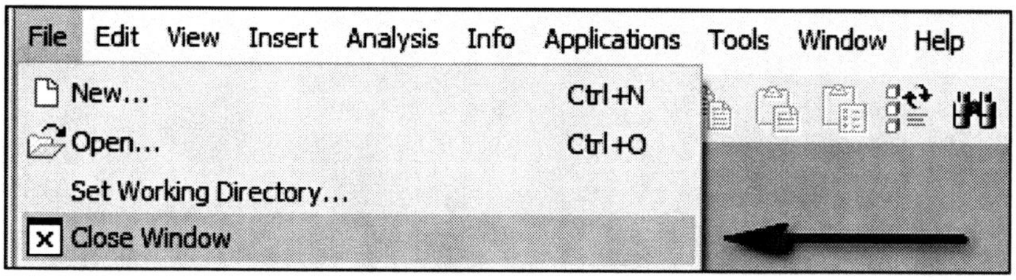

Option 2:

Open PE4. Select **File** > **Set Working Directory** as shown in Figure 2-2.

Figure 2-2

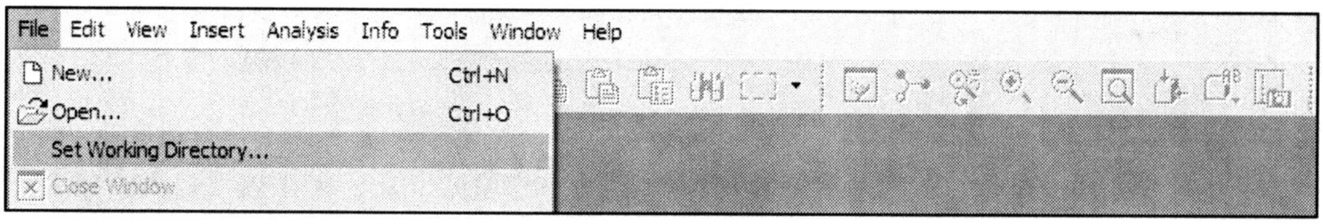

Set the working directory to the folder that the instructor has indicated.

As with the battery cover, select **File** > **New** (see Figure 2-3) or use the shortcut to create a new part.

Figure 2-3

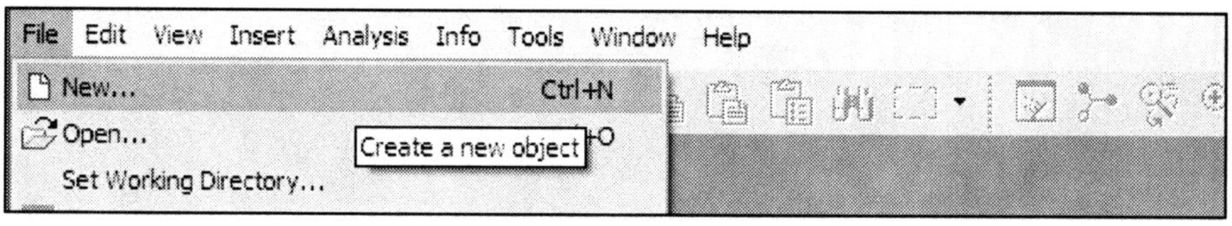

Name the part, "Battery" and select **OK** in the **New** dialog box.

Make sure that the PE4 default unit set **(Inch Ibm Sec)** is used. Follow the instructions in Part 1 (pp. 15-16) if you do not remember how to check and change the units.

As before, a series of basic sketches and extrusions is used to form the 3-D battery geometry. If you are confused during the next steps, refer to the more detailed instructions associated with making the battery cover in Part 1.

Select the **TOP** datum plane and choose the **Sketch** icon. Select **Sketch** in the **Sketch** dialog box.

Select the **Create circle** icon located in the vertical toolbar on the right shown in Figure 2-4.

Figure 2-4

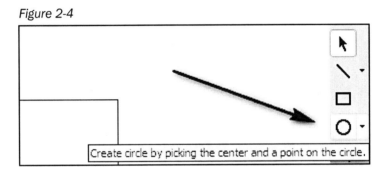

Create circle by picking the center and a point on the circle.

Point the mouse over the intersection of the dashed lines at the center of the screen (the **Front** and **Right** datum references). When on center, the small red circle will appear indicating that an "on-point" constraint will be activated.

Once the constraint appears, left-click to begin drawing the circle. Just as with the rectangles you sketched in Part 1, move the mouse away from the center to expand the circle. Left-click again to complete the circle.

After the circle is completed, hit the **Select** (pointer) icon at the top of the vertical toolbar. A dimension will appear indicating the diameter of the newly sketched circle. Edit this dimension to **.55**.

Select the **Check Mark** to complete the sketch.

When returned to the 3-D environment, select the sketch and then pick the **Extrude** tool (see Figure 2-5).

Figure 2-5

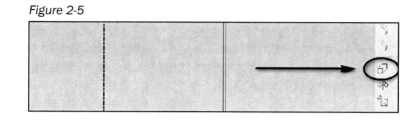

Edit the extrusion depth to a value of **1.8** (see Figure 2-6). Select the **Green Check** to complete the extrusion (see Figure 2-7).

Figure 2-6

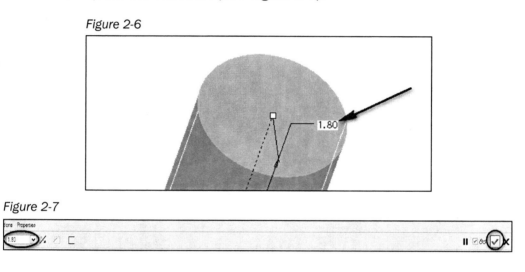

Figure 2-7

You should now have a basic cylinder.

In PE4 there are two basic methods for creating an extrusion. A sketch can be drawn followed by the use of the **Extrude** tool (as done earlier) or the **Extrude** tool can be chosen first with an **internal sketch** definition following. The latter is attempted here.

Use the **Selection Filter** at the bottom of the screen to activate the **Geometry** filter. Select either of the two flat cylinder faces (see Figure 2-8).

Figure 2-8

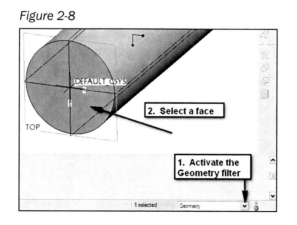

Select the **Extrude** tool.

In the 3-D modeling area, hold down the right mouse button to activate the drop-down menu. Let go of the right mouse button and left-click on the **Define Internal Sketch** option (see Figure 2-9).

Figure 2-9

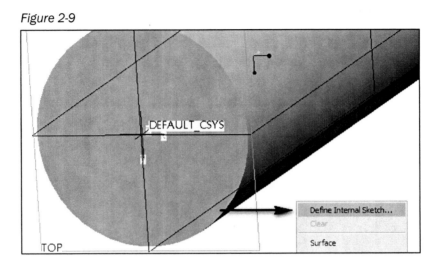

Select **Sketch** in the **Sketch** dialog box to begin the next drawing.

Draw another circle, again starting at the intersection of the dashed lines. Make the diameter small enough to fit inside the perimeter of the cylinder face. After the circle has been drawn, select the **Pointer** to activate the diameter dimension. Change the dimension to **.2** (see Figure 2-10).

Figure 2-10

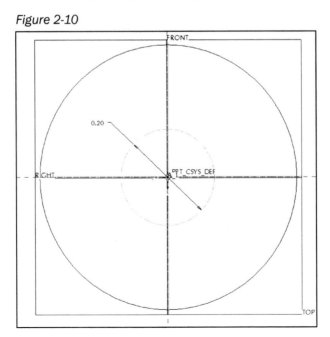

Select the **Check Mark** to complete the sketch.

Now, edit the depth of the extrusion to a value of **.1** using either of two methods described earlier (see Figure 2.11). Once this is complete, select the **Green Check** as always to complete the extrusion.

Figure 2-11

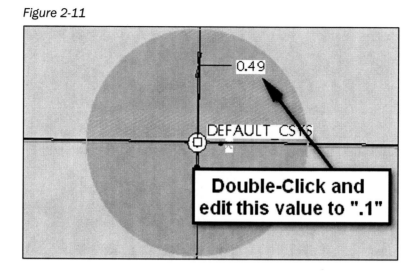

Reorient the figure so that both of the extrusions can be seen clearly. So far, the battery should look like the image shown in Figure 2-12.

Figure 2-12

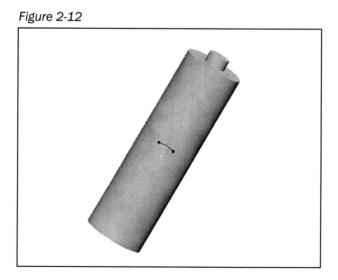

Next, the focus shifts to giving some more realistic characteristics to the top and bottom edges of the cylinders. First, the middle edge needs to be rounded off.

Again, activate the **Geometry** filter at the bottom of the screen.

Select the middle edge as indicated in Figure 2-13.

Figure 2-13

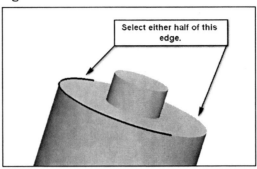

Select either half of this edge.

Pick the **Round** tool from the vertical toolbar on the right shown in Figure 2-14.

Figure 2-14

Round Tool

Notice that PE4 automatically generates a preview with a dimension available for editing, just as it does with extrusions. Change this dimension to a value of **.05** (see Figure 2-15). This dimension controls the radius of the round. Thus, by setting it to .05, the round begins and ends on the flat and curved surfaces .05 inches from the edge.

Figure 2-15

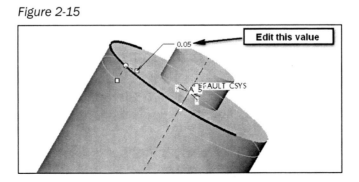

0.05   Edit this value

DEFAULT CSYS

Also as with extrusions, pick the **Green Check** to complete the round. You can also hit the middle button of your mouse to indicate completion.

Apply these same steps to the small, top edge of the battery. This time use a value of **.02** for the round radius.

The result should look like Figure 2-16.

Figure 2-16

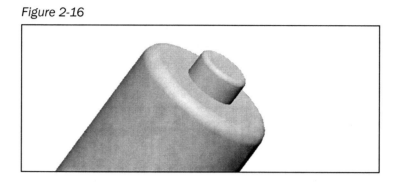

For the last edge a chamfer is used. A chamfer is an angled cut that is commonly used to eliminate sharp, 90 degree edges from steel parts. The steps required to apply a chamfer are very similar to those required for a round. However, the chamfer has a few more options. The **Chamfer** tool is located directly below the **Round** tool.

Select the bottom edge followed by the **Chamfer** tool (see Figure 2-17).

Figure 2-17

Activate the drop-down chamfer menu in the upper left. Notice the variety of options available for creating a dimension. All of these options are various ways of creating the angle of the chamfer.

Select the **D1 x D2** option (see Figure 2-18). This, like the round, will control the start and end points of the chamfer.

Figure 2-18

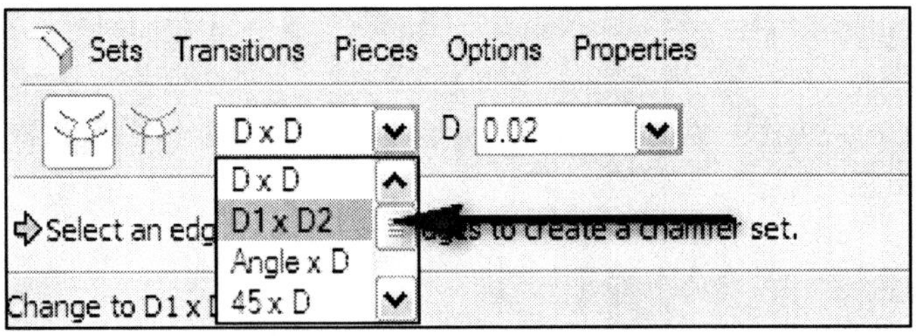

Assign the value for **D1** as **.1** and that for **D2** as **.05** (see Figure 2-19). This causes the chamfer to start .1 inches from the edge on the flat face and finish .05 inches up the curved surface of the cylinder.

Figure 2-19

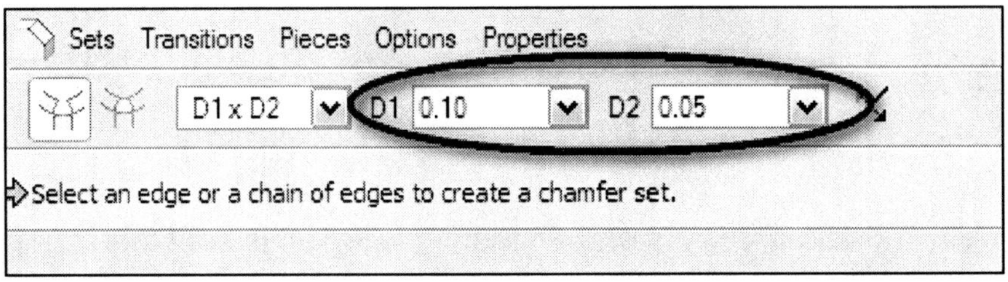

Again, select the **Green Check** or **middle-click** to complete the chamfer. The final battery should appear like this.

Figure 2-20

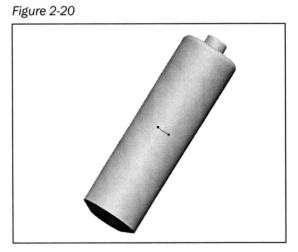

Select **File** > **Save** or use **Control + S** to save the part.

Part 2a is complete!

In Part 2b an identical battery is created using a revolution about a central axis. The revolution will require more sketch work, but will eliminate the need for extrusions.

Part 2b—Battery (2)

If continuing directly from Part 2a (without closing PE4), read the following instructions, but refrain from taking action until you create the new part (*).

Open the battery that was just created so that it can be referenced to make sure this second (but identical) battery has the same dimensions.

First, make sure to **set the working directory** to the folder with your previously made part files (the battery cover and the battery).

There are 3 ways to open a file in PE4.
1. Select **File** > **Open** (see Figure 2-21)
2. Pick the **Open** icon on the left side of the top horizontal toolbar (see Figure 2-22)
3. Press **Control + O**

Figure 2-21

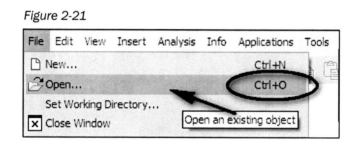

Figure 2-22

*Create a new part file. Name it "Battery_2." Take note of the use of the underscore "_" since PE4 will not allow the use of a space between "Battery" and "2."

Again, check to make sure that PE4 default units **(Inch lbm Sec)** are being used. If PE4 is consistently starting up in the default unit set, do not worry about this. However, if the units are consistently wrong, it is imperative to change them every time a new part is created. Errors become very evident in the assembly phase as complications arise in trying to place a 2 inch battery into a 5 mm casing.

First, select the **Revolve** tool located just below the **Extrude** tool (see Figure 2-23).

Figure 2-23

Next, hold down the right mouse button and select **Define Internal Sketch** as done in Part 2a.

Select the **TOP** datum plane and start a sketch.

Begin the sketch by constructing a vertical centerline along the dashed line representing the **RIGHT** datum plane (see Figure 2-24; refer to Part 1 if help is needed).

Figure 2-24

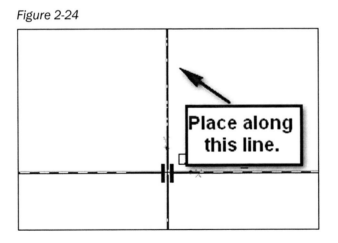

This centerline will become the central axis around which the sketch will be revolved in order to form the 3-D geometry. This sketch will represent a cross-section of the part (battery) that was just made.

Change from the **Centerline** to **Line** tool (see Figure 2-25) and start drawing at the central intersection of the dashed lines.

Figure 2-25

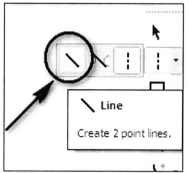

Left-click to start the line and move it horizontally to the right about an inch on the screen along the **FRONT** datum reference. **Left-click** again to end the line segment. Notice that as the first segment is ended, a second one is automatically started. Cancel the creation of the second segment by clicking the **middle mouse button**.

Select the **Pointer** to activate the dimension. The first battery was created with a diameter of .55 inches. The base segment should be half this length since it is a radius. PE4 can perform simple calculations within the sketcher. **Double-click** the dimension box to edit it and enter **.55/2** in the box.

PE4 automatically carries out the division and changes the dimension to a value of **.275**.

*NOTE: PE4 might round up to **.28**. If this happens, **double-click** the value. PE4 will display **.275**. Hit enter confirming that value. PE4 will occasionally round up from thousandths to hundredths place decimals.*

NOTE: Depending on how PE4 adjusts, you may need to zoom in very far to see the adjusted line.

The next step is to start another line segment where the previous segment ended. Select the **Line** tool again (see Figure 2-26) and move over the end of the segment. Notice that the mouse pointer goes straight to the red circle when it is "sucked in" by the end point. Once this happens, click to start the line.

Figure 2-26

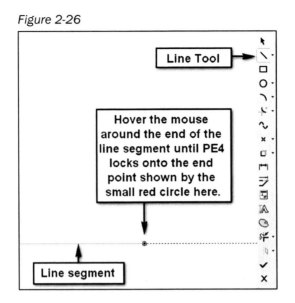

Draw a diagonal segment and continue drawing a shape like the figure shown in Figure 2-27. **Middle-click** to end the line series after the final horizontal segment reaching the centerline has been formed. Do not worry about specifying dimensions. This is only a rough sketch.

If a mistake is made, terminate the line series with a **middle-click**. Then, select the mistakes and press **Delete** on the keyboard.

Figure 2-27

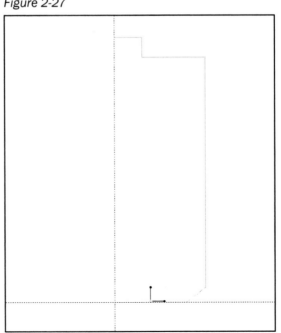

Once the basic sketch is created, click the **Dimension** icon in the vertical toolbar on the right (see Figure 2-28).

Figure 2-28

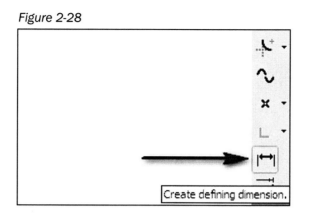

Zooming in with the **middle mouse wheel**, focus in on the diagonal line segment at the base of the sketch. Notice that the line is currently defined by a dimension and an angle. Calculating the angle would require trigonometry. Instead PE4 can be manipulated to display vertical and horizontal dimensions.

Select the top point of the diagonal segment and follow by clicking on the dashed **FRONT** reference line. **Middle-click** between the two selection spots to activate a vertical dimension (see Figure 2-29).

Figure 2-29

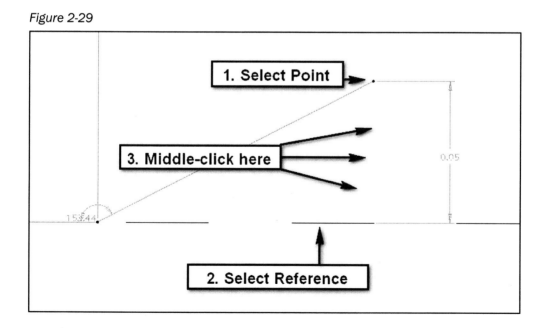

Choose **OK** in the **Select** dialog box (Figure 2-30).

Figure 2-30

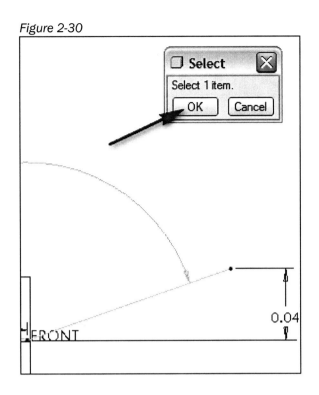

Double-click and edit the new dimension to a value of **.05**.

Left-click on the **Dimension** icon again. This time, select the points on either end of the segment by clicking each once. Then, **middle-click** in the area above the top end point and inside the rest of the sketch (see Figure 2-31).

Figure 2-31

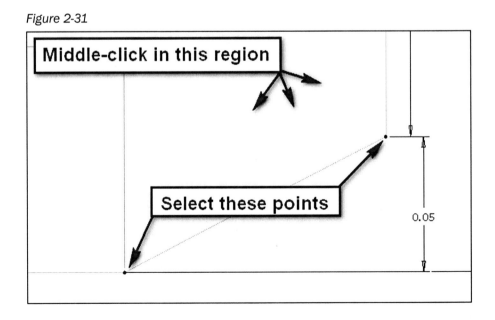

NOTE: Dimensions can be positioned almost anywhere in the sketcher by clicking and dragging them to another part of the screen.

Again, choose **OK** in the **Select** dialog box and edit the dimension to **.1**.

The **.1** and **.05** values stem from the choices for **D1** and **D2** that were made in creating the chamfer in Part 2a.

Notice that the remaining four line segments in the image below have an **H** or **V** next to them. These are constraints that force the line to be **Horizontal** or **Vertical**. When sketching lines, pay attention to these constraints as they may or may not be desirable.

Figure 2-32

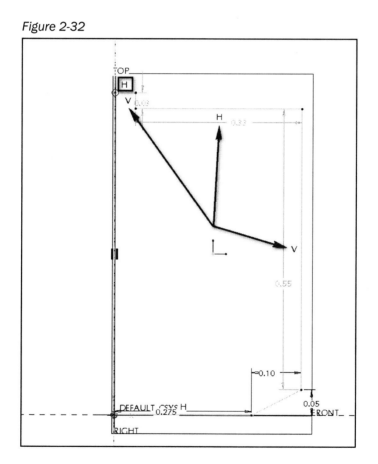

For example, if you were trying to roughly sketch a line at 87 degrees, PE4 might form a vertical constraint that will block you from easily editing the angle of the line since it was set to 90 degrees (vertical) by the system. If one is ever formed by accident, simply select the constraint and press the **Delete** key as done with the equal distance constraints appearing in Part 1.

In this case, the constraints are desirable. If you are missing any of the constraints shown in the above image, select the **Constrain** tool in the vertical toolbar on the right (see Figure 2-33).

Figure 2-33

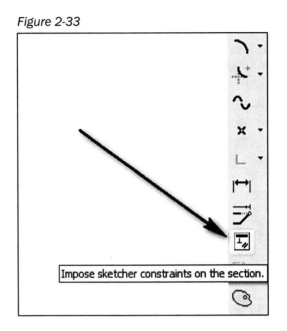

The top left icon in the **Constraints** dialog box is for **vertical constraints** and the one next to it is for **horizontal constraints**.

To apply one of these two constraints, select the appropriate icon (**vertical** or **horizontal**, see Figure 2-34) and then select the line. Then choose **OK** in the **Select** dialog box or pick another constraint icon to apply to the sketch.

Figure 2-34

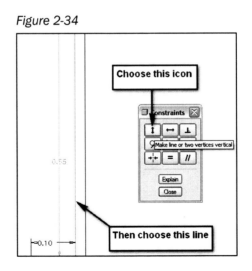

In some cases, it is possible to over-constrain a line segment. PE4 will force the choice between constraints producing an error message like the one shown in Figure 2-35.

Figure 2-35

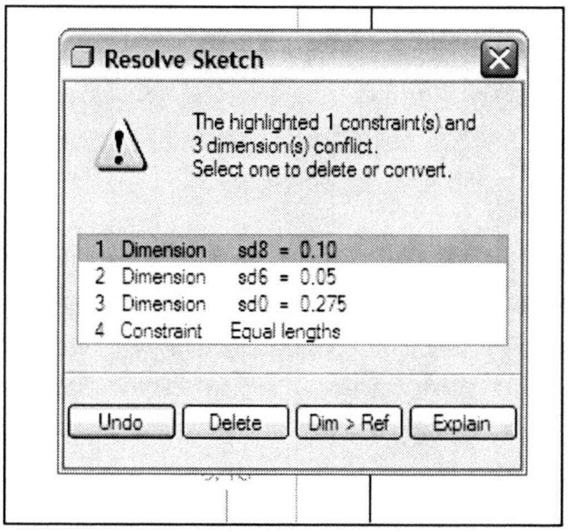

```
┌─────────────────────────────────────────┐
│  □ Resolve Sketch                    ⊠   │
│                                          │
│        ⚠    The highlighted 1 constraint(s) and │
│             3 dimension(s) conflict.     │
│             Select one to delete or convert. │
│                                          │
│   1  Dimension    sd8 = 0.10             │
│   2  Dimension    sd6 = 0.05             │
│   3  Dimension    sd0 = 0.275            │
│   4  Constraint   Equal lengths          │
│                                          │
│   [ Undo ]  [ Delete ]  [ Dim > Ref ]  [ Explain ] │
└─────────────────────────────────────────┘
```

Selecting **Undo** in the **Resolve Sketch** dialog box will remove the constraint that was just placed. The other option is to delete an existing constraint. As you mature in PE4, you will gain the wisdom to know which ones to select and delete. For now, if the **Resolve Sketch** dialog box appears, choose **Undo**.

When the four constraints have been activated, select **Close** in the **Constraints** dialog box.

Edit the three faint dimensions indicated in Figure 2-36 to the following values.
1. **1.75**
2. **.175**
3. **.1**

Figure 2-36

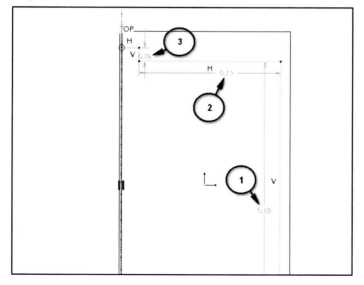

When entering the second dimension, PE4 may round the number to the hundredths decimal place (.18). If this happens, enter **.175** again and the exact number will appear.

If three dimensions are not available, click the **Dimension** tool. Select the line that is missing the dimension and then move the mouse just off the line and **middle-click**. The dimension should appear.

Notice that applying these remaining three dimensions will automatically constrain the last segment (the top horizontal segment). An attempt to call up a dimension using the technique just explained will yield the **Resolve Sketch** dialog box. Try this and see. The dimension would be redundant so choose **Undo**.

Your result should look like Figure 2-37.

Figure 2-37

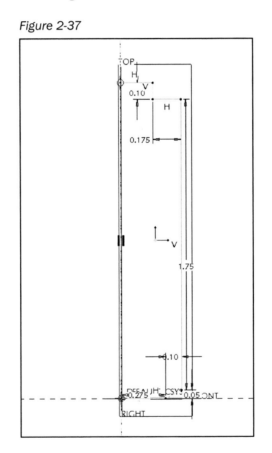

Select the **Check Mark** to complete the sketch.

PE4 prompts that the **"Section is Incomplete"** (see Figure 2-38). The message area indicates that the sketch needs to be closed. These messages will be very useful as you try using more advanced tools.

Figure 2-38

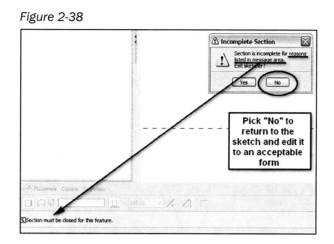

Select **NO** and return to the sketch. The centerline that was started with does not count as a perimeter line in the sketch. An additional vertical line needs to be drawn along this centerline connecting the top and bottom line segments of the sketch. Select the **Line** tool and close the sketch. Make sure that the **On Point** (red circle) constraints pop up as they did in the creation of the diagonal segment.

Choose the **Check Mark** to complete the sketch again.

Notice the advantage gained when using an internal sketch to create revolutions. Had the sketch been made before activating the **Revolve** tool, PE4 would not have known what the sketch was for and thus could not have given the warning that the sketch was unacceptable for a revolution.

PE4 automatically generates a 360 degree revolution (see Figure 2-39). If only a partial revolution is desired, the degrees of rotation can be altered in the text box in the upper left. **Middle-click** or pick the **Green Check** to complete the full 360 degree revolution.

Figure 2-39

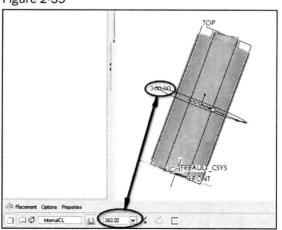

Notice how this battery looks different than the first one; the diameter is disproportionately large.

This diameter can be measured using the analysis tools built into PE4.

Select **Analysis** > **Measure** > **Diameter** (see Figure 2-40).

Figure 2-40

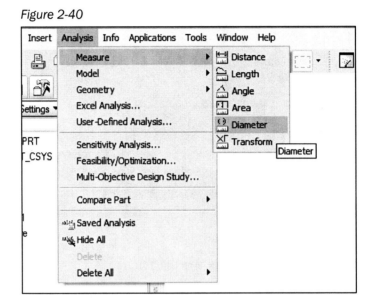

With the **Type** set to **Quick**, select the curved surface of the battery (see Figure 2-41).

Figure 2-41

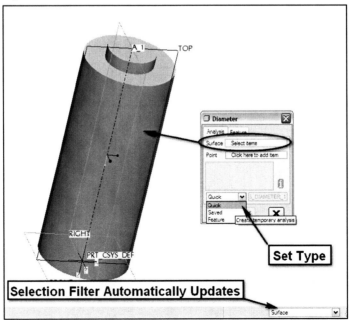

Notice that the **Diameter** dialog box indicates a value of **.75** inches. The next step is to check the diameter on the first battery. Select the **Green Check** in the **Diameter** dialog box (Figure 2-42) to complete the analysis.

Figure 2-42

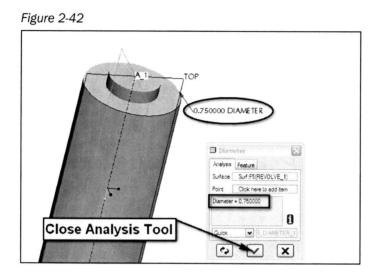

Move to the bottom of your screen and locate the window labeled "Battery."

Notice that the current window, "Battery_2" has "(Active)" next to it. PE4 can have multiple parts open at one time, but only one can be active. In order to perform the same diameter analysis on the original part "Battery" the window must be activated.

Figure 2-43

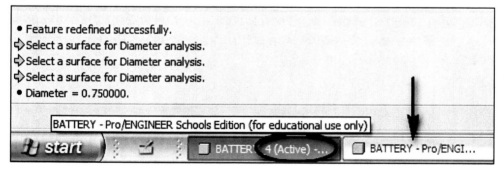

Select the window (BATTERY) indicated by the arrow in Figure 2-43.

Notice that as the mouse is moved around the screen, a black circle with a line through it is present indicating the lack of activation.

Press the keys **Control + A** in combination. This **activates** the window (see Figure 2-44).

Figure 2-44

◻ BATTERY **(Active)** - Pro/ENGINEER Schools Edition (for educational use only)

Check the diameter of this original battery.

The analysis indicates a value of **.55** inches.

Go back to the window containing "Battery_2" and reactivate the part using **Control + A**.

The first diameter is .2 larger than the second. Decreasing the first segment drawn in the revolution sketch by .1 should achieve the desired shrunken result.

Expand the item labeled **Revolve 1** in the **Model Tree** on the left by selecting the small **+** (see Figure 2-45).

Figure 2-45

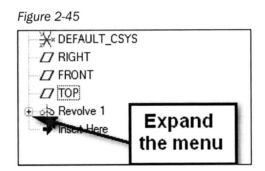

S2D0001 (the sketch) appears under **Revolve 1** because **S2D0001** is the parent to the child, **Revolve 1**. **Right-click** on **S2D0001** and select **Edit Definition** from the pop-up menu (see Figure 2-46).

Figure 2-46

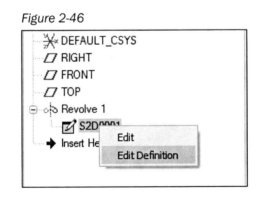

Zoom in on the base of the sketch rolling the **mouse wheel backwards**. Edit the first segment's value from **.275** to **.175** (see Figure 2-47).

Figure 2-47

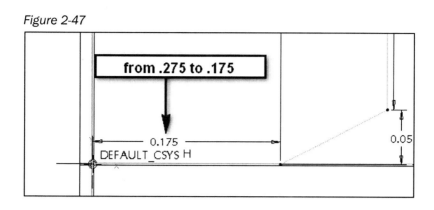

Notice how all the segments move over by .1 horizontally since they are all interconnected and the length of one of them has changed.

Select the **Check Mark** to complete the sketch. The model regenerates.

The new battery looks much more like the last one. Now all that is required are the two rounds. The rounds also could have been included in the base sketch by adding arcs to the top two corners, but in the interest of time and simplicity, the rounds are applied after the revolution has been completed.

Using the **Round** tool as in Part 2a, place a **.02** inch radius round on the top edge and a **.05** inch radius round on the other (middle) unfinished edge.

Press **Control + D** to change the view to a standard orientation. Press **Control + S** and save the part. The result is shown in Figure 2-48.

Figure 2-48

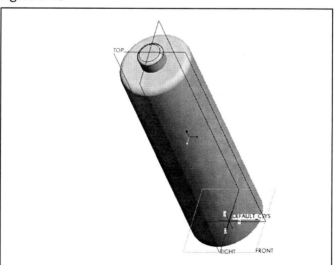

Part 2b of this tutorial is now complete. Next is the circuit board for the device.

Part 3—Circuit Board

*Remember to set the working directory and check that the units are set to Pro/E Default.

Start a new part file. Name it "Circuit_Board."

Start a new sketch on the **TOP** datum plane.

Draw a **.7 x 2** inch rectangle like the one shown in Figure 3-1. Start at the intersection of the reference lines. For help, refer to the rectangle drawings in Part 1.

Figure 3-1

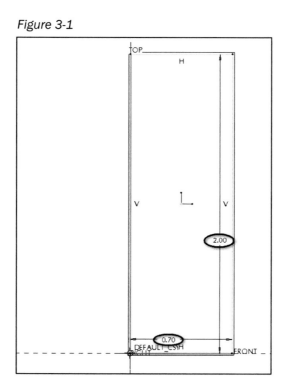

Complete the sketch and extrude it to a depth of **.15** (see Figure 3-2).

Select **Extrude 1** in the **Model Tree** and change the selection filter to **Datums** (see Figures 3-3, 3-4).

Figure 3-2

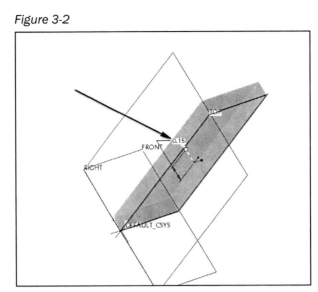

Figure 3-3

Figure 3-4

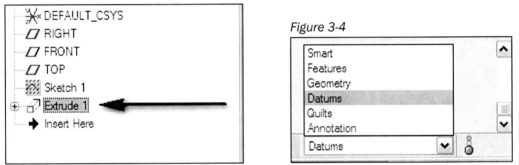

Select the **Mirror** tool located in the vertical toolbar on the right shown in Figure 3-5.

Figure 3-5

The object to be mirrored has already been selected (**Extrude 1**). The message area at the top indicates all that remains to be done is the selection of the plane across which the extrusion will be mirrored. Select the **RIGHT** datum plane. Be careful when making the selection (see Figure 3-6).

Figure 3-6

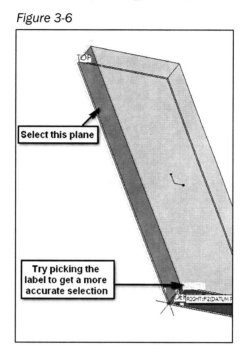

Middle-click or pick the **Green Check** to complete the mirror process. Notice in Figure 3-7 that the extrusion is now twice as wide.

Figure 3-7

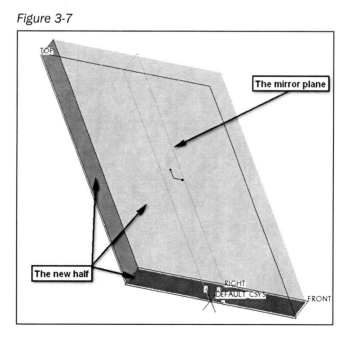

Set the **Selection Filter** to **Geometry** and choose the top surface of the part (see Figure 3-8).

Figure 3-8

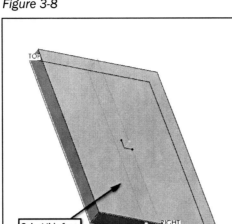

Enter the sketching environment.

In the menu at the top of the sketcher, select **Sketch** > **References** (see Figure 3-9). This aids in the creation of extra reference lines that will be useful when defining the dimensions of the sketch.

Figure 3-9

Select the top and right edges of the part to form reference lines on as indicated in Figure 3-9.

More dashed lines (references) appear as in Figure 3-10.

Figure 3-10

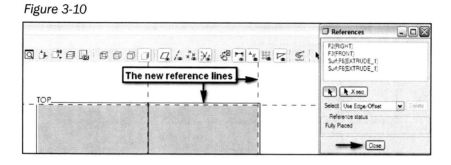

Select **Close** in the **References** dialog box.

Go to the **3-Point Arc** tool and select the small arrow next to it to activate the menu (see Figure 3-11). Choose the **Arc Center** tool.

Figure 3-11

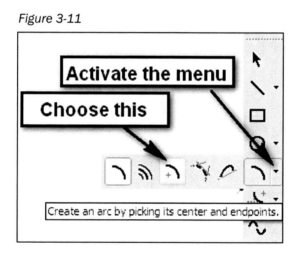

The **Arc Center** tool can create any portion of the circumference of any circle. The radius of the circle is determined by how far the start point is from the center point. For this part (Circuit Board), the tool is used to create semi-circles.

Move the start point on top of the dashed line running down the center of the part (the **RIGHT** datum reference; see Figure 3-12). Center the arc near the top of the part by clicking on the reference line.

Start the arc by selecting just above the center point on the reference line. Finish the arc by moving in a clockwise direction to below the center of the arc and selecting the reference line again. A clockwise motion just ensures that the semi-circle will be on the right hand side of the reference line.

Figure 3-12

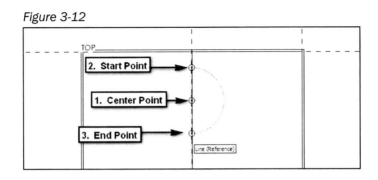

Select the **Dimension** tool. Click the center point of the arc and follow by clicking the top reference line. **Middle-click** inside the part and to the left of the **RIGHT** datum reference (see Figure 3-13).

Figure 3-13

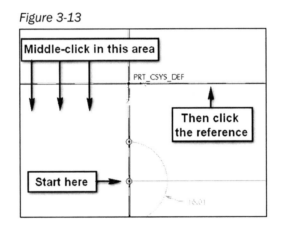

Choose **OK** in the **Select** dialog box.

Edit the new dimension to a value of **.3**. Change the radius value to **.25** (see Figure 3-14).

Figure 3-14

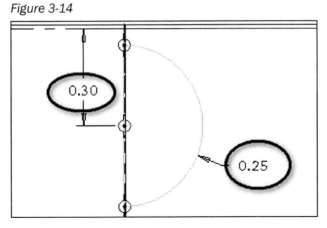

Create a similar arc using the **Arc Center** tool. Place this semi-circle near the bottom of the part face. Take care not to accidentally constrain the second arc to have the same radius as the first. This will appear as an **R1** symbol. If it appears after the arc is drawn, select it and delete it.

Call up the vertical dimension as before. Choose **OK** in the **Select** dialog box and edit the dimension to a value of **1.6**. Edit the radius to **.2** (see Figure 3-15).

Figure 3-15

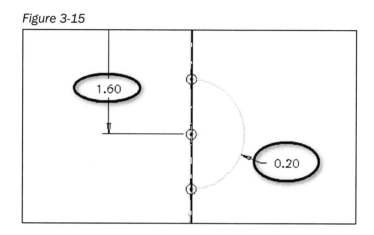

Lastly, create a circle to the right of the central reference line and between the two arcs. Again, be careful not to create any accidental constraints.

This time, two dimensions will be necessary to completely define the planar position of the circle since it is not already constrained to the centerline.

Use the **Dimension** tool and select the center of the circle and the top reference line followed by a middle-click to get the vertical dimension. Change the dimension to **1.1**. Choose **OK** in the **Select** box as before.

Repeat the process instead using the right most reference line (the other of the two that were added). **Middle-click** to the right of the part to complete the dimension. End the selections.

Edit the new value to **.25**. Change the diameter value to **.4** (see Figure 3-16).

Before the sketch is completed, close the two arcs by adding line segments along the centerline (as with the revolution sketch in Part 2b). Start at the bottom point and end at the top (see Figure 3-17).

Figure 3-16   *Figure 3-17*

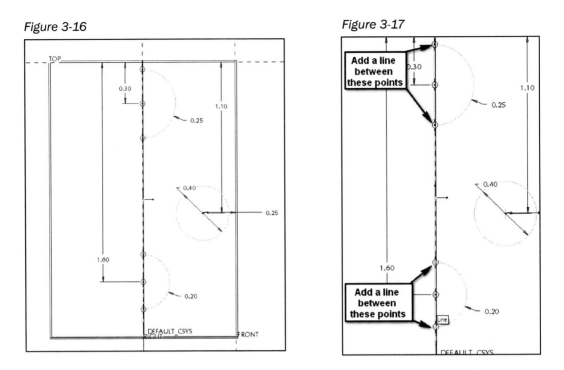

Complete the sketch and extrude it to a depth of **.3** (see Figure 3-18).

Figure 3-18

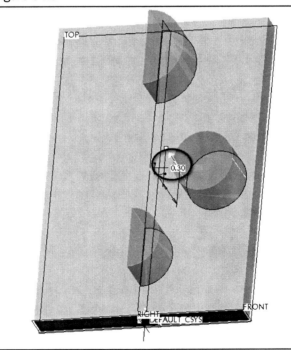

If necessary, select **Extrude 2**. Just as was done with **Extrude 1**, mirror the second extrusion about the **RIGHT** datum plane (see Figure 3-19).

Figure 3-19

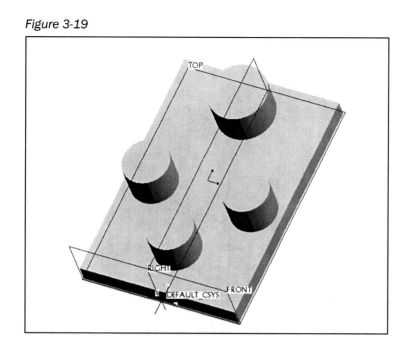

Change the selection filter to **Geometry**.

Using the **Round** tool, round the 4 top edges of the cylinders. All of the edges can be rounded at once by first selecting one edge and holding down the **Control** key and picking the remaining three edges (see Figure 3-20). Make the round radius a value of **.1** and complete the feature.

Figure 3-20

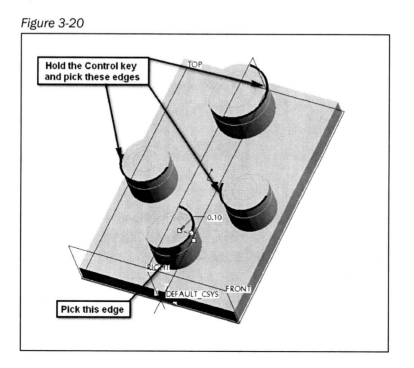

Select the top face of the part as in Figure 3-21.

Figure 3-21

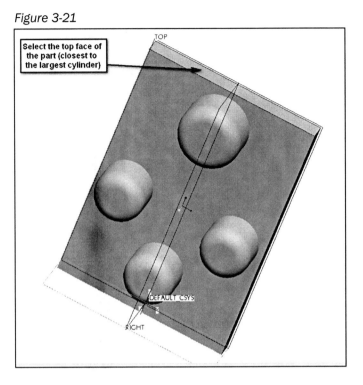

Select the **New Datum** tool near the top of the vertical toolbar on the right (see Figure 3-22).

Figure 3-22

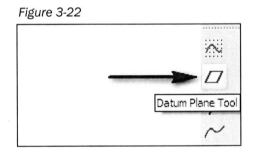

In the **Datum Plane** dialog box, enter **-.3** for the value of the translation as in Figure 3-23. The negative is required to get the plane to shift in the opposite direction of the yellow arrow. Choose **OK**.

Figure 3-23

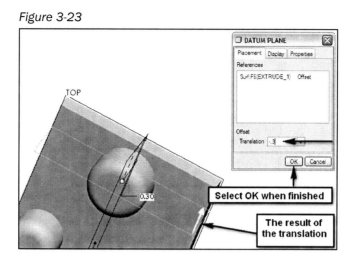

Go to the **Model Tree** and locate the newly created datum plane at the bottom.

Right click the new plane and select **Rename** (see Figure 3-24). Input "Plane1" as the name and press **Enter** on the keyboard.

Figure 3-24

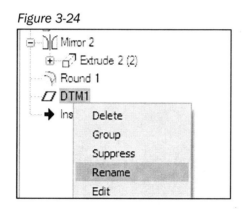

Next, a datum axis will be created. It is important to realize that the intersection of two planes forms a line. Thus, two intersecting datum planes can be used to completely define the placement of a datum axis since it will be forced to lie on that line.

Go to the **Model Tree** and select the **Right** datum plane. Hold down the **Control** key and select **Plane1**.

Select the **Axis** tool (see Figure 3-25).

Figure 3-25

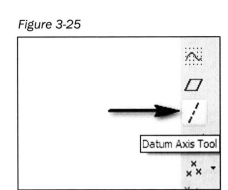

The axis automatically appears (see Figure 3-26), already constrained by the previous selection. These two steps can be performed in reverse order and will yield the same result.

Figure 3-26

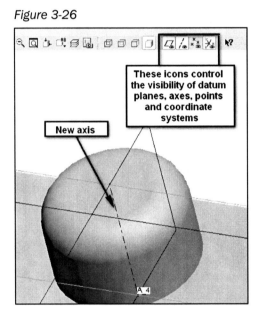

Note: The axis visibility must be active. Make sure that the visibility icon indicated in Figure 3-26 is highlighted (pressed) at the top of the screen.

Now that the buttons are complete, the largest button will be distinguished by covering it in tiny bumps that could be recognized through touch.

Select **Plane1** and start a new sketch (see Figure 3-27).

Figure 3-27

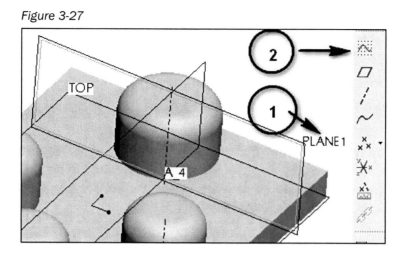

The sketch is oriented in an awkward position. Notice that the **Sketch** dialog box has **"Right"** in the box next to **Orientation**. If an awkward orientation appears like this in the future, look for the orientation and select an edge that you want to appear on that side of the final orientation. For example, selecting the edge indicated in Figure 3-28 will cause that edge to appear on the right side of the screen. Select the edge and observe the result.

Figure 3-28

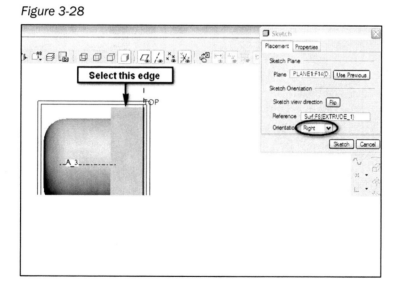

Select **Sketch** in the **Sketch** dialog box to continue with the newly oriented sketch.

Select **Sketch** > **References** in the top menu.

Select the top edge of the middle button. Also select the new datum axis located under **Plane1** near the base of the model tree (see Figure 3-29). **Close** the **Reference** dialog box when finished.

Figure 3-29

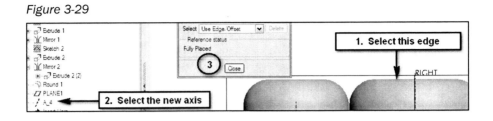

Zoom in on the intersection of the reference lines that were just created at the center of the middle button. Use the **Arc Center** tool (see Figure 3-30) to create a quarter of a circle centered at the intersection and running from the vertical to the horizontal reference line. Make the radius **.02**.

Figure 3-30

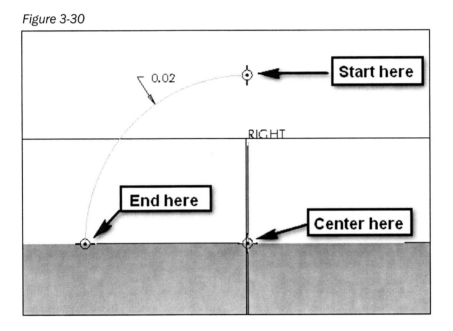

Complete the sketch.

Select the **Revolve** tool (see Figure 3-31). The revolution needs an axis. Select the new axis (see Figure 3-32).

Figure 3-31

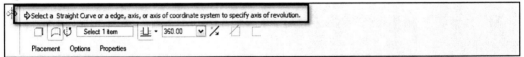

Figure 3-32

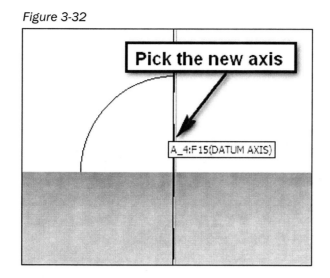

Make sure that the revolution is a full 360 degrees. **Complete** the revolution.

Select **Revolve 1** now in the model tree.

Select the **Pattern** tool at the bottom of the vertical toolbar on the right (see Figure 3-33). The **Pattern** tool allows the user to replicate features on a large scale.

Figure 3-33

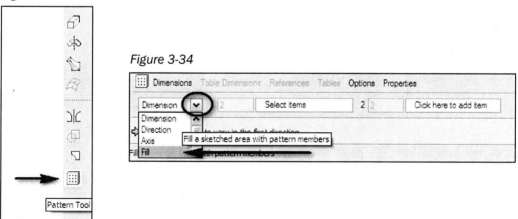

Figure 3-34

Select **Fill** (see Figure 3-34) from the drop down list in the upper left corner.

In the model view area, hold down the right mouse button and select **Define Internal Sketch** from the pop-up menu.

Select the top surface of the largest button (see Figure 3-35) where the revolution feature was just created.

Figure 3-35

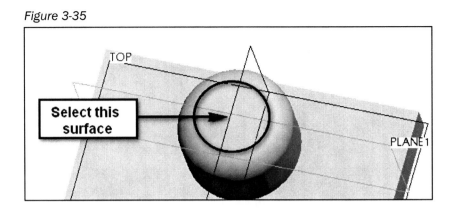

Select **Sketch** in the dialog box to continue.

Add in a reference line along **Plane1**. Pick **Sketch** > **References** > **Plane1**. **Close** the **References** dialog box.

Sketch a circle placing the origin at the intersection of the **Plane1** and central reference lines. Give the circle a diameter of **.3** as shown in Figure 3-36.

Figure 3-36

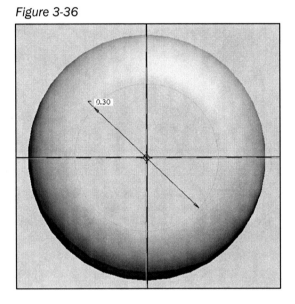

Complete the sketch.

Look back to the top left and edit the separation between entities to a distance of **.05** and as shown in Figure 3-37. Press **Enter** on the keyboard. The default pattern is a square, but the choice is yours. Notice how the black dots appear on the button showing where each of the replicated features will be placed (see Figure 3-38).

Figure 3-37

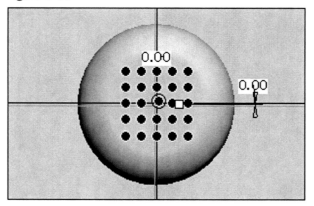

Figure 3-38

The pattern is completed like all other features; **middle-click** or pick the **Green Check**.

To get a better view of the new features, deactivate the **axis visibility**. This will also speed up 3-D rotations of the part. It is important to remember to reactivate the visibility when finished as visibility options will carry over from part to part.

Save the part. You have completed the circuit board. On to the top of the shell.

Part 4—Cap

Set the working directory and check the units.

Create a new part. Name it "Cap."

Begin a new sketch on the **Top** datum plane.

Start with the **Ellipse** tool shown in Figure 4-1.

Figure 4-1

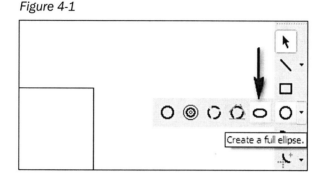

Start at the intersection of the references and move the mouse away from the point to expand the ellipse as with the circle. Attempt to get an egg shape like the one shown in Figure 4-2.

Figure 4-2

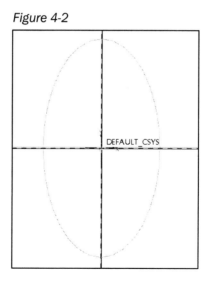

Select the **Pointer** to pull up the dimensions. Change them as shown in Figure 4-3. Depending on what dimensions are started with, the sketch may become flattened. Just focus in on different areas and find the dimensions to edit them.

    1. $R_x = 1.5$ 2. $R_y = 2.5$

Figure 4-3

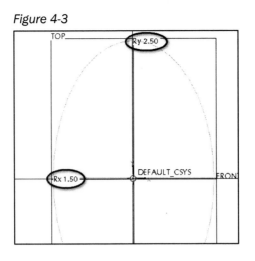

Select the **Dynamic Trim** tool shown in Figure 4-4.

Figure 4-4

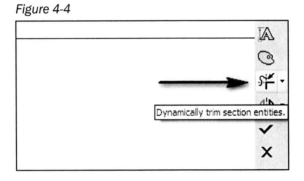

Select the bottom two arcs to eliminate them from the sketch (see Figure 4-5).

Figure 4-5

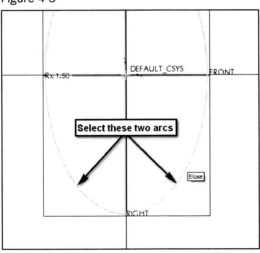

Choose the **Arc Center** tool and finish the bottom of the sketch with a half circle (see Figure 4-6).

Be sure to activate the **On Point** constraints. If they are not active, a dimension will appear for the radius of the half circle when the **Pointer** is selected afterwards.

Figure 4-6

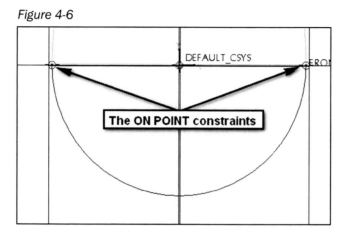

Complete the sketch and **extrude** it to a depth of **.5** inches.

Apply a round of radius **.25** to the top edge (see Figure 4-7).

Figure 4-7

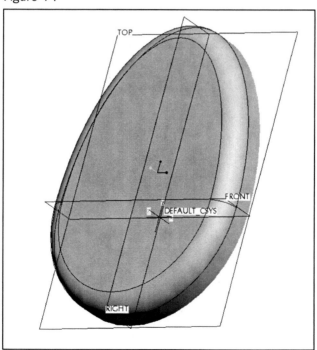

Set the selection filter to **Geometry.**

Roll the part over and select the flat, bottom surface indicated in Figure 4-8.

Figure 4-8

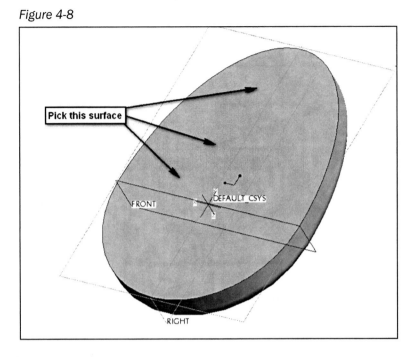

Select the **Shell** tool (see Figure 4-9) in the vertical toolbar on the right.

Figure 4-9

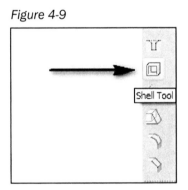

There is a text box in the upper left that controls the **Thickness** of the shell. Make this a value of **.1** (see Figure 4-10).

Figure 4-10

🔲 References Options Properties
Thickness .1 ✓ ⚒

Complete the shell (see Figure 4-11).

Figure 4-11

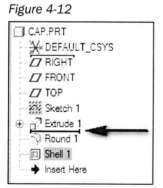

Note that if the round were applied after the shell, the resulting figure would have differed. This can be examined simply.

Go to the **Model Tree**. Click and drag **Shell1** above **Round1**. The thin black bar (see Figure 4-12) indicates where the feature is being inserted in the model tree.

Figure 4-12

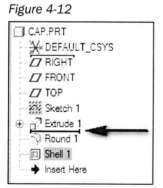

Notice how there is no longer internal curvature on the part (see Figure 4-13).

Figure 4-13

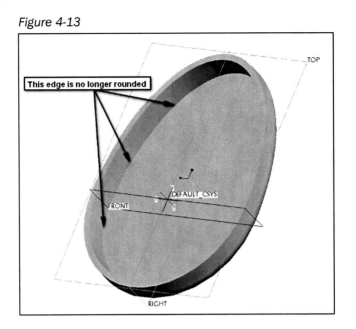

Drag **Shell1** below **Round1** to its original position.

Select **Insert** > **Sweep** > **Cut** as shown in Figure 4-14.

Figure 4-14

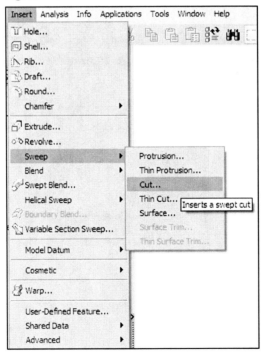

Notice the **CUT: Sweep** dialog box in the upper right. A cut requires three elements to be defined. The first element is the **Trajectory**, the path of the cut, which will be defined shortly. The second is the **Section**, which is the 2-D shape that will be dragged along the trajectory to form the cut. Finally, the **Material Side** is used to determine whether material will be left above or below the trajectory.

As one element's definition is completed, the next definition automatically begins until the **message area** indicates that all elements have been defined.

Select **Sketch Trajectory** from the **Menu Manager** sub menu (see Figure 4-15).

Select **Plane** and then choose the **TOP** datum plane.

If the red arrow is facing the wrong direction (see Figure 4-16), choose **Flip** then **Okay**. Otherwise, just choose **Okay**. Flip changes the orientation of the sketch in advance.

Figure 4-15

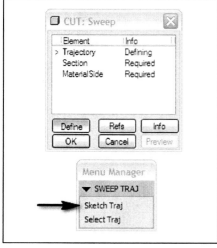

Figure 4-16

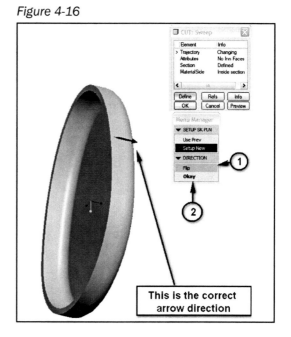

Choose **Default**.

Select the **Offset Edge** tool (see Figure 4-17) in the vertical toolbar on the right.

Figure 4-17

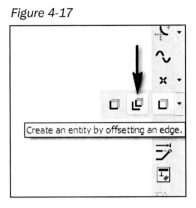

Set the **Type** to **Chain**.

Select the top and bottom outside edges (see Figure 4-18). These are separate edges because they were made with separate segments earlier.

Figure 4-18

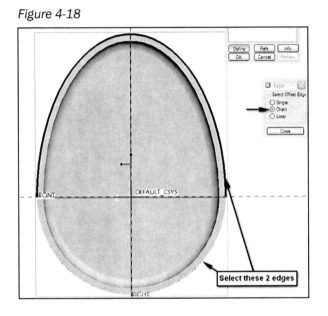

If the **Convert Chain to Loop** dialog box appears, select **No** (as in Figure 4-19).

Figure 4-19

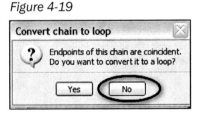

The **message area** prompts for an offset entry in direction of the arrow. Enter **-.05** and press the **Enter** key.

A line is created similar to the first sketch made for this part. Select **Close** under **Type**.

Complete the sketch.

Pick **No Inn Fcs** > **Done**.

Focus in on the intersection of the **datums**. Starting at the intersection, draw a **.04 x .1** rectangle (see Figure 4-20).

Figure 4-20

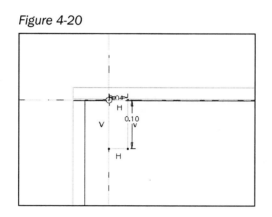

Complete the sketch.

Select **Okay** in the **Menu Manager**.

The **message area** (see Figure 4-21) now indicates that all of the elements needed have been defined.

Figure 4-21

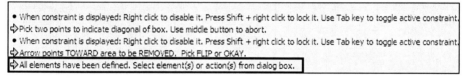

Select **OK** in the **CUT: Sweep** dialog box.

The cut is made, but notice that there is unwanted material leftover (see Figure 4-22).

Figure 4-22

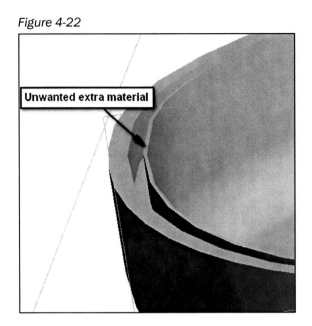

The **Section** was not wide enough. Locate the cut in the model tree. Right click it and select **Edit Definition** (see Figure 4-23).

Figure 4-23

Select **Section** in the **CUT: Sweep** dialog box. Follow by picking **Define** (see Figure 4-24).

Figure 4-24

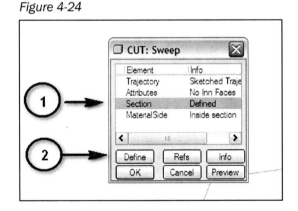

Change the **.04** dimension to **.06**. **Complete** the sketch.

All of the other elements maintain their previous definition. Select **OK** under **CUT: Sweep**.

The new cut eliminates the extra material (see Figure 4-25).

Figure 4-25

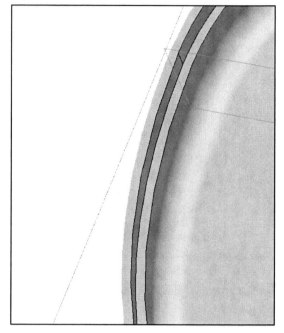

The final step is to cut holes in the top so the buttons on the circuit board can come through enough to be pressed.

With the selection filter set to **Geometry**, choose the top surface (see Figure 4-26) and start a new sketch.

Figure 4-26

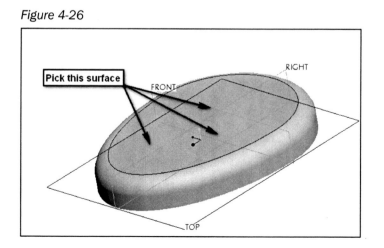

Make sure that the part is oriented with the wider part at the top like Figure 4-27.

Figure 4-27

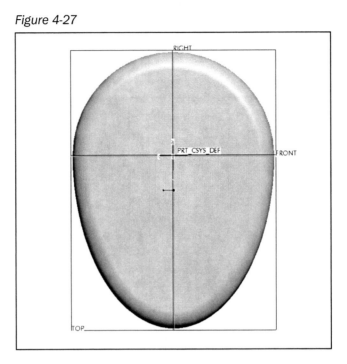

Create a sketch with four circles placed according to the dimensions displayed in Figure 4-28.

Figure 4-28

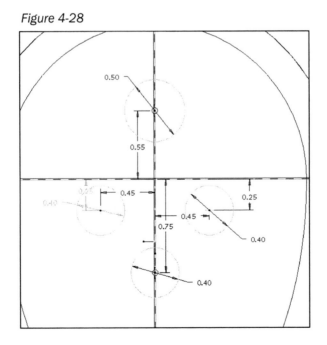

The following subsection will take a step-by-step approach if this is too difficult.

*Start by creating a circle of diameter **.5** on the center vertical reference line. The center should be **.55** inches above the horizontal reference line (see Figure 4-29). For help placing dimensions, refer back to the detailed instructions in Part 3.*

Figure 4-29

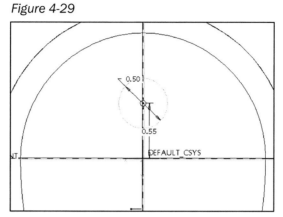

*The second circle is of diameter **.4**. It sits on the same centerline, but below the horizontal by **.75** inches (see Figure 4-30). Watch out for the R1 symbol that can occur when drawing the circle, indicating equal radii between the first circle and the second. Delete the R1 to remove the constraint and edit the diameter.*

Figure 4-30

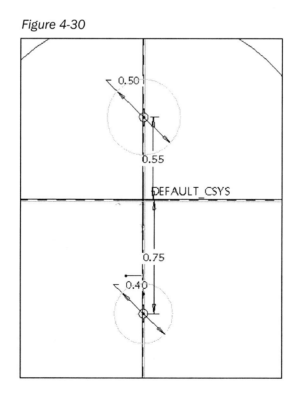

*Draw a third circle of diameter **.4** a distance of **.45** right of the vertical and **.25** below the horizontal. In this case, the R1 is acceptable if applied to the lower circle (see Figures 4-31, 4-32, 4-33).*

Figure 4-32

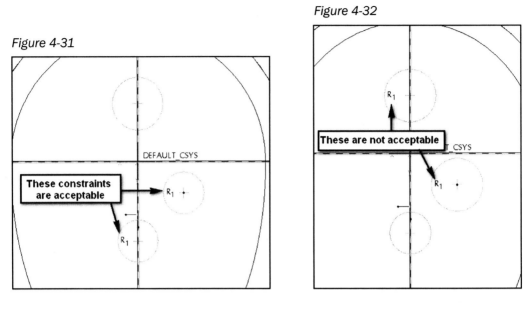

Figure 4-31

Figure 4-33

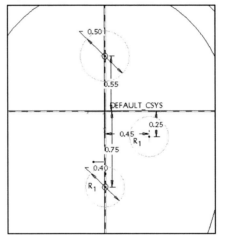

*The fourth circle is the same as the third except it is **.45** left of the vertical. There is another constraint option that appears before the circle is started. Notice the red dash indicating that the fourth circle will be placed level with the third. Starting the circle while this dash appears will eliminate the need to set the vertical dimension on the fourth circle. Also, try to activate the same R1 constraint to save time. This way the .45 is all that needs to be manually entered (see Figures 4-34, 4-35).*

Figure 4-34

Figure 4-35

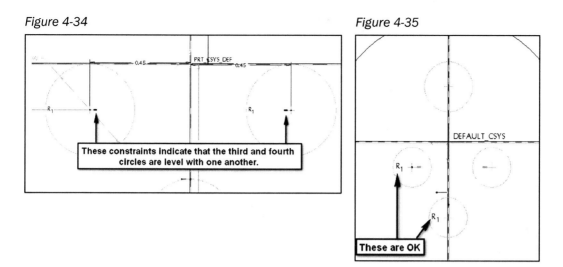

Complete the sketch.

To finish the part, the new sketch will be negatively extruded to form holes.

Select the **Extrude** tool.

Select **Remove Material** icon (see Figure 4-36) from the toolbar in the upper left. Also, select the **Extrude to Depth** option from the list.

Figure 4-36

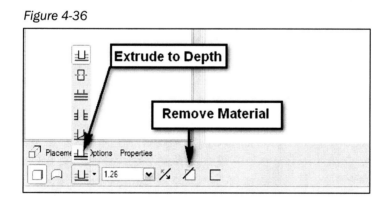

The **Remove Material** option is what switches the extrusion to a negative extrusion.

The **Extrude to Depth** option requires a surface to extrude to (from the surface that was sketched on).

Rotate the part to expose the inside and select the surface indicated in Figure 4-37.

Figure 4-37

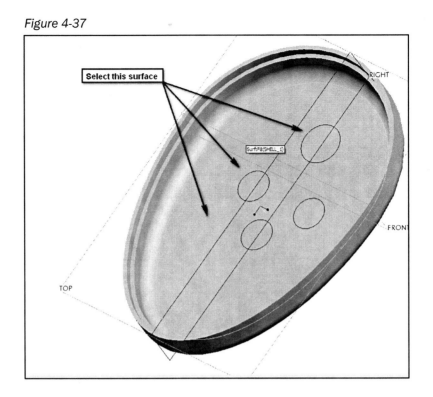

If unable to select, choose the **Select Item** box shown in Figure 4-38.

Figure 4-38

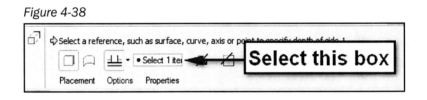

Complete the extrusion just like all the others.

Instead of using the **Extrude to Depth** option, the direction of the extrusion could have been changed and the depth could have been set to **.1**. **Extrude to Depth** is more effective if the thickness of the part is not readily known.

Save the part.

Close the window.

Part 4 is now complete. The bottom half of the outer shell will be created next.

Part 5—Base

Set the working directory and check the units.

Create a new part. Name it "Base."

It is important to pay close attention to the directions for this part as it will be the biggest challenge yet. Figure 5-1 is an image of the finished product.

Figure 5-1

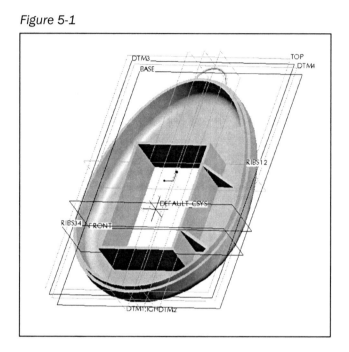

Select **Insert** > **Sweep** > **Protrusion** (see Figure 5-2).

Figure 5-2

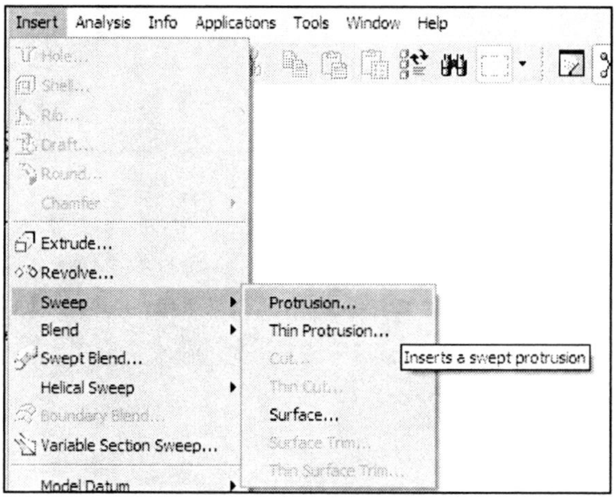

This sweep will function similar to the cut, except it will form 3D geometry instead of remove it.

Pick **Sketch Trajectory**. Select **Plane** and pick the **TOP** datum plane.

Select **OK** > **Default.**

Sketch the combination of ellipse and half circle used in Part 4. The sketch is shown in Figure 5-3 for reference.

Complete the sketch.

Select **Add Inn Fcs** > **Done.**

The needed sketch is shown in Figure 5-4. It is broken into steps to ensure accuracy.

Figure 5-3   *Figure 5-4*

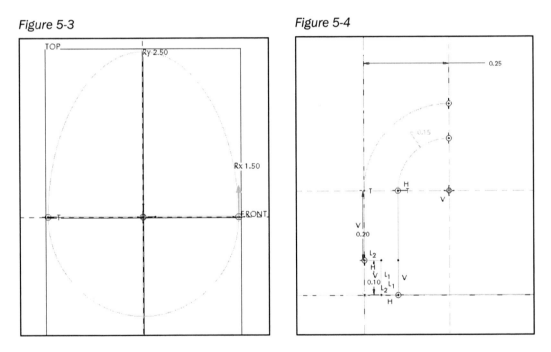

Start by creating 2 centerlines; one vertical and one horizontal. Place the vertical centerline **.25** inch to the right of the current vertical reference. Place the horizontal **.3** inch above the horizontal **(TOP)** reference (see Figure 5-5).

Figure 5-5

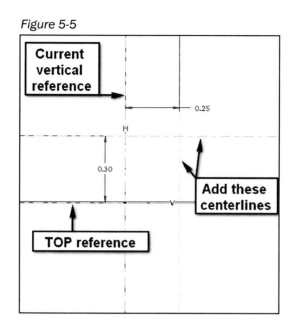

Draw a line segment from the **TOP** reference to the horizontal **centerline** .1 inch right of the vertical **reference**, shown in Figure 5-6.

Figure 5-6

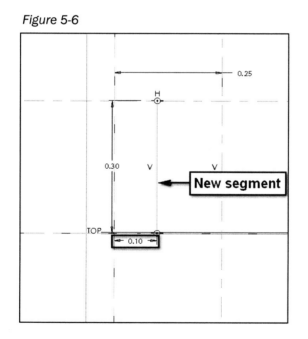

Next, use the **Arc Center** tool to create a quarter circle centered at the intersection of the **centerlines**. Start the arc at the top end of the line segment and finish on the **vertical centerline**. Create another similar quarter circle centered at the same intersection. This arc should run from the intersection of the **vertical reference** and **horizontal centerline** up to the **vertical centerline**. The result is shown in Figure 5-7.

Figure 5-7

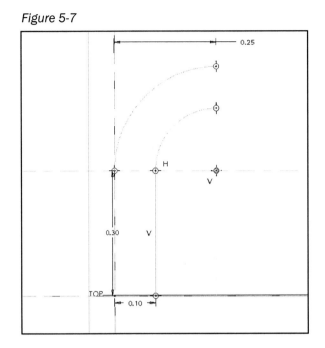

Notice that the arc radii are not required as they are dictated by the centerlines.

Start a line segment at the bottom of the larger arc (see Figure 5-8). Draw the segment along the **vertical reference**. Set the distance to **.2** inch.

Figure 5-8

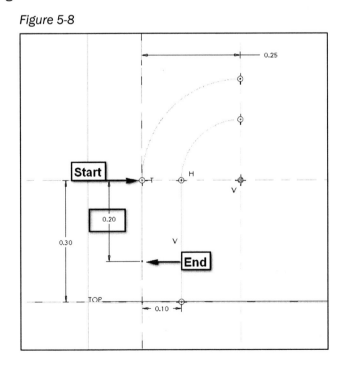

Follow up with three short segments to close the sketch (see Figure 5-9). The first horizontal segment is the only one that requires a dimension. Set it to **.05** inch.

Figure 5-9

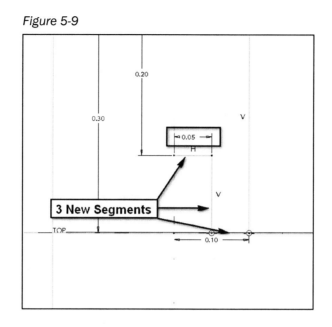

*Note: The sketch is left open due to the **Add Inn Fcs** (Add Inner Faces) option that was selected earlier. Material will be bridged from one part of the trajectory to the other to form one solid figure instead of the ring that would be formed by closing the figure and choosing **No Inn Fcs**.*

Complete the sketch.

Again, the **message area** indicates that all of the necessary elements have been defined. Select **OK** in the **PROTRUSION: Sweep** dialog box or **middle-click** to complete the feature.

The result is essentially the "Cap" without holes (see Figure 5-10). The most important distinction is that the material was removed to the right of the ridge as opposed to the left in the "Cap." This was done so that the "Cap" would fit over the "Bottom."

Figure 5-10

With the selection filter set to **Geometry**, start a new sketch on the large inner surface indicated in Figure 5-11.

Figure 5-11

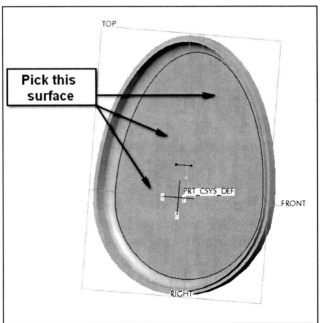

PE4 automatically activates the **References** dialog box since there are not currently enough references to properly constrain a sketch.

Select the **Front** datum to create a reference (see Figure 5-12). Select **Close** in the **References** dialog box.

Figure 5-12

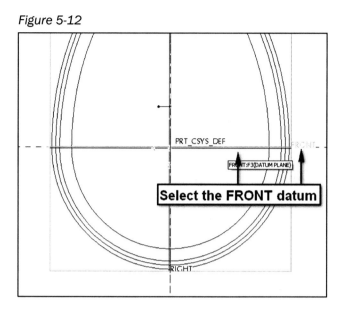

Draw a **2** x **1.4** inch rectangle (see Figure 5-13). The bottom line of the rectangle should be **.9** inch below the **TOP reference** line that was just created. Also, the rectangle should be symmetric about the **RIGHT vertical reference** line. To accomplish this, dimension either of the vertical sides of the rectangle **.7** inch off of the **vertical reference** line.

Figure 5-13

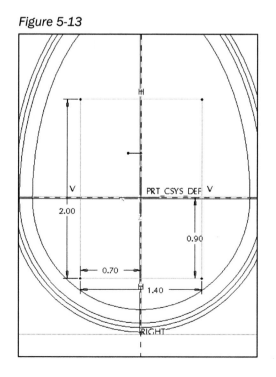

Complete the sketch. **Extrude** the sketch **.6** inch.

Create a new sketch on the top face of the new extrusion.

Looking directly on the surface, the extrusion seems to have disappeared. Spin the part to see the extrusion. Reset the orientation by choosing **View** > **Orientation** > **Sketch** from the top menu of the sketcher.

Select the **Offset Edge** tool that was used in Part 4. Set the **Type** to **Chain**. Select the right and top edges of the extrusion. PE4 automatically completes the rest of the loop (highlighted in red). Choose **Accept** in the **Menu Manager** and **NO** in the **Convert Chain to Loop** dialog box.

Enter a value of **-.05** for the offset value (use +.05 if the orange arrow points toward the center of the part).

Select **Close** under **Type**. The result is shown in Figure 5-14.

Figure 5-14

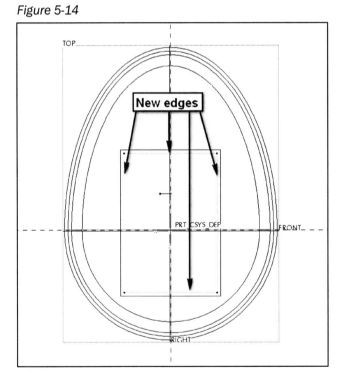

Complete the sketch.

Select the **Extrude** tool. Choose the **Remove Material** option. Also select the **Extrude to Depth** option (see Figure 5-15).

Figure 5-15

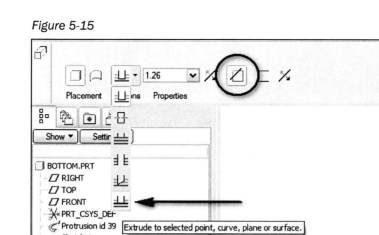

Select the flat surface on the bottom of the part shown in Figure 5-16.

Figure 5-16

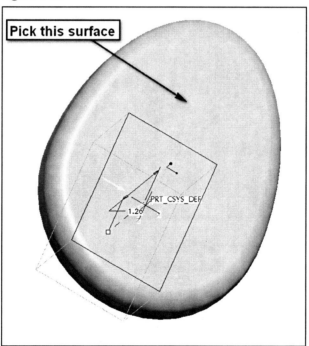

Complete the negative extrusion.

Set the selection filter to **Geometry**. Start a new sketch on the broad surface that was just extruded to. Again, make sure that the widest portion is oriented toward the base of the screen in order to have synergy between your sketch and the one described here.

Once more, PE4 prompts the user to add references to the sketcher. Select the

RIGHT and **FRONT** datum planes (see Figure 5-17). **Close** the **References** dialog box.

Choose the **Offset Edge** tool. Set the **Type** to **Chain.** Select the right and top edges of the hole.

Choose **Next** followed by **Accept** in the **Menu Manager**. Choose **NO** in the **Convert Chain to Loop** dialog box.

If the yellow arrow points toward the center of the part, enter a value of **-.05** in the text box at the top of the screen (otherwise, use +.05). Choose **Close** in the **Type** dialog box.

Figure 5-17

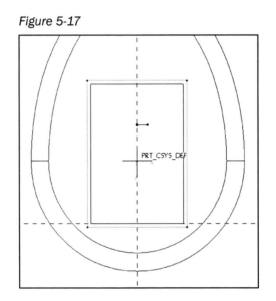

Create a small **.6** x **.3** inch rectangle along the top line of the larger rectangle just formed (see Figure 5-18). The small rectangle should be symmetrical around the **RIGHT** datum reference.

Figure 5-18

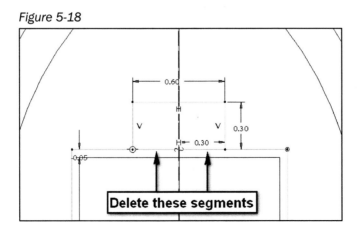

As with the "Battery Cover," delete the redundant lines between the two rectangles with the **Dynamic Trim** tool.

Complete the sketch. **Extrude** a distance of **.05** into the part with the **Remove Material** option active (see Figure 5-19).

Figure 5-19

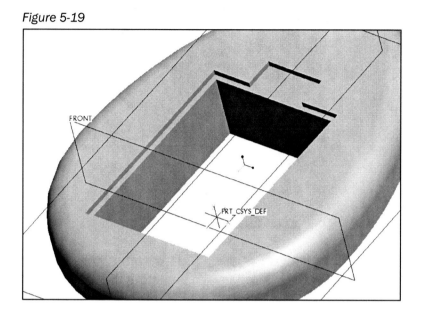

The remainder of the part will require some intensive work with datum planes. Some of this work is necessary for the part's features, while the others will be needed in completing the final assembly.

Select the **RIGHT** datum plane followed by the **Datum Plane** tool in the vertical toolbar (see Figure 5-20).

Figure 5-20

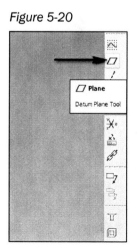

Enter a value of **.3** into the text box in the **Datum Plane** dialog box (see Figure 5-21). **Middle-click** to complete the feature or select **OK** in the **Datum Plane** dialog box.

Figure 5-21

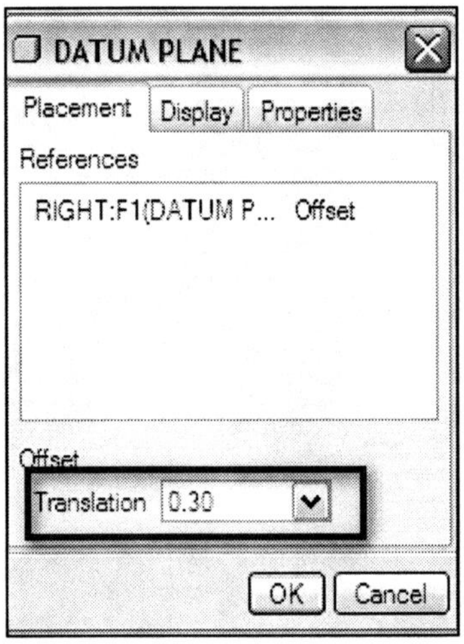

DATUM PLANE

Placement | Display | Properties

References

RIGHT:F1{DATUM P... Offset

Offset

Translation | 0.30

OK | Cancel

Create another datum plane off of the **RIGHT** datum plane. This time, enter a value of **-.3** into the offset text box before completing the feature.

The two new datum planes should appear like those shown in Figure 5-22.

Figure 5-22

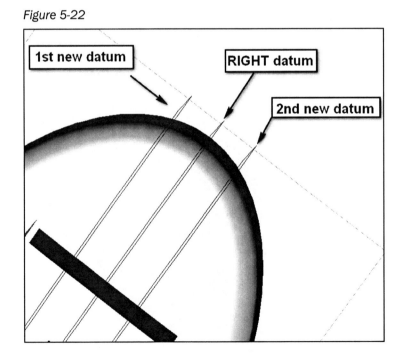

1st new datum

RIGHT datum

2nd new datum

With the **Selection Filter** set to **Geometry**, select the flat, large inside surface of the part shown in Figure 5-23.

Figure 5-23

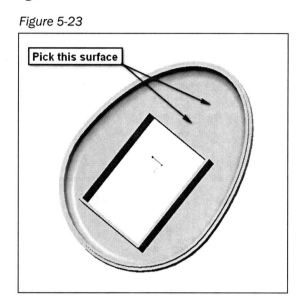

Pick this surface

Create a new datum plane with an offset of **-.3**. This new plane should pass through the rectangular geometry.

Make sure that the new datum plane (**DTM3**) is still selected. Holding the **Control** key, select **DTM1**. Select the **Datum Axis** tool (below the **Datum Plane** tool). The axis is created automatically just as in Part 3.

Repeat this process using **DTM2** in place of **DTM1** to create a new separate axis (see Figure 5-24).

Figure 5-24

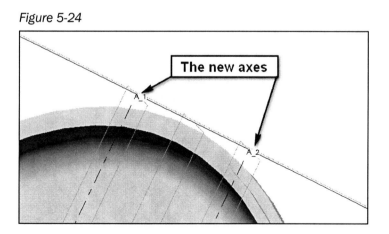

The new axes

In the **Model Tree** select **A_1.** Right-click the object and select **Rename** from the drop-down list. Rename the axis to **X1**. Press **Enter** on the keyboard. Rename **A_2** as **X2**.

Create another datum plane off of the same surface used to create **DTM3**. Use an offset of **-.2**. Rename this plane as **RING.** Then, create another datum plane off of the same surface with **0** offset. Rename this last plane as **BASE.**

Select the **FRONT** datum plane. Create a new plane off of the **FRONT** plane with an offset of **.75**. It should be between **FRONT** and the thin part of the egg shape. Otherwise, use **-.75**. Rename the plane to **RIBS12**.

Select the vertical wall at the bottom of the part shown in Figure 5-25 and offset another plane by **-.4**. Name it **RIBS34**.

Figure 5-25

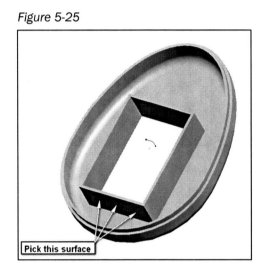

Select **RIBS12**.

Orient the part on your screen like Figure 5-26. This will ensure consistency between your screen and the upcoming images.

Figure 5-26

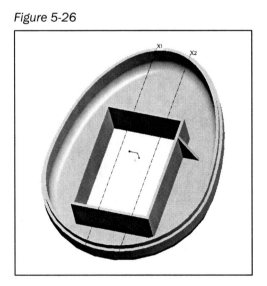

Start a new sketch.

Select **Sketch** > **References** and select the 2 edges and the **BASE** datum plane shown in Figure 5-27.

Figure 5-27

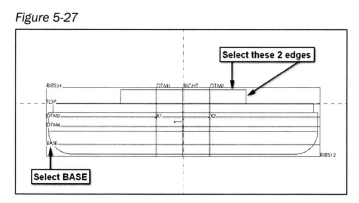

Start a line where the horizontal and vertical edges meet (see Figure 5-28). Extend the line diagonally downward until the reference line formed by selecting **BASE** is reached. Set the angle to 65 degrees (use 25 degrees if the angle is dimensioned from the other side of the line).

Figure 5-28

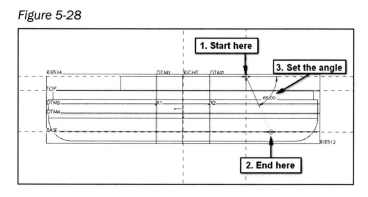

Complete the sketch.

Select the **Rib** tool in the vertical toolbar shown in Figure 5-29.

Figure 5-29

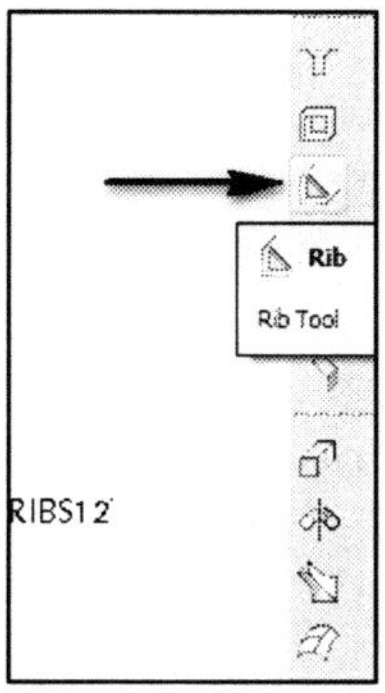

The **Rib** tool operates off of an open sketch. It effectively fills in the area between the line and the surfaces touched by the line's endpoints.

It is imperative that the direction of the rib be corrected or else the feature will not work. The yellow arrow appearing in the part demonstrates the direction of the fill. This should be pointing inward toward the part. The direction is correct when the shaded yellow area appears in the image (see Figures 5-30, 5-31).

Figure 5-30

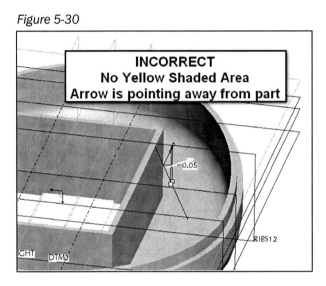

Figure 5-31

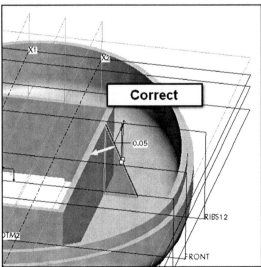

The width of the rib can be set numerically by editing the value in the text box at the upper right. Also, the number appearing on the feature can be edited as with an extrusion or round (see Figure 5-32).

Figure 5-32

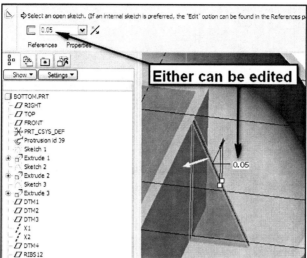

Change the width to **.05**.

The arrow symbol next to the text box controls the side of the line that the rib is created on. It can now either run .05 inch to the right or left of the line. It can also run symmetrical about the line. Use this symmetry option. This is indicated by the presence of two white boxes in the modeling area shown in Figure 5-33.

Figure 5-33

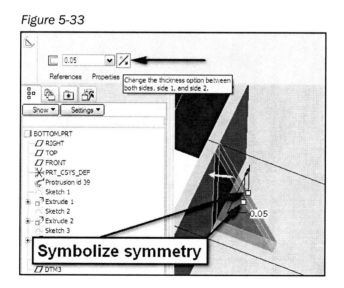

Next, create a new rib opposite this one.

Select **RIBS12** and reorient the part as you did before.

Complete a sketch like the one shown in Figure 5-34.

Figure 5-34

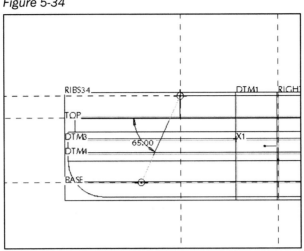

Make an identical rib.

Repeat this process using the plane **RIBS34**.

The result is shown in Figure 5-35.

Figure 5-35

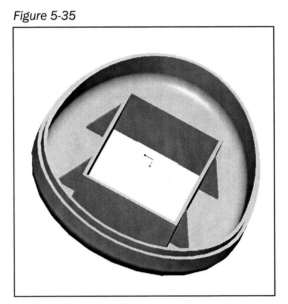

Next, a key ring will be added to the shell as this remote device is like those used to unlock a car. This will be done with a swept protrusion.

Select **Insert** > **Sweep** > **Protrusion**.

Choose **Sketch Trajectory** in the **Menu Manager** and then select the datum plane labeled **RING**. Select **Okay, Default** in the **Menu Manager**.

The sketcher prompts for the addition of references. Select the outermost edge of the part shown here as well **FRONT, RIGHT, DTM1,** and **DTM2** to form reference lines (see Figure 5-36). **Close** the **References** box.

Figure 5-36

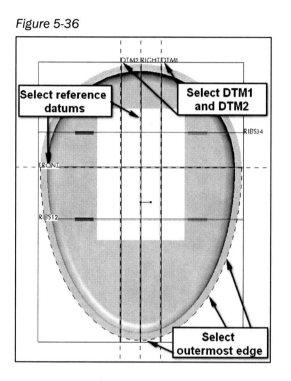

Select the **Arc Center** tool. Place the center point at the intersection of the **RIGHT** datum and the outer edge reference lines.

Start the arc at the intersection of **DTM1** and the outer edge reference line. Finish at the intersection of **DTM2** and the outer edge (see Figure 5-37).

Figure 5-37

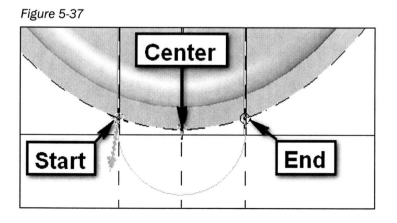

Complete the sketch.

Select **Merge Ends** > **Done** in the **Menu Manager**.

Draw a **.03** diameter circle as the section. Remember to center it on the intersection of the dashed lines.

Complete the sketch.

Select **OK** in the **Sweep: Protrusion** box or **middle-click** to complete the feature.

The result is shown in Figure 5-38.

Figure 5-38

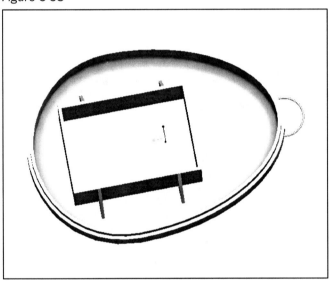

Finally, one more hole will be cut so that the battery cover can fit into the bottom shell. Select the surface indicated in Figure 5-39.

Figure 5-39

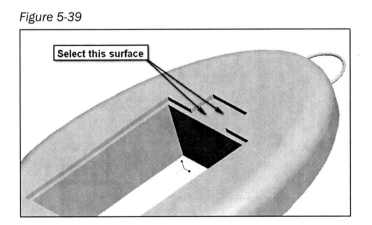

Start a new sketch.

Create the reference lines shown in Figure 5-40.

Figure 5-40

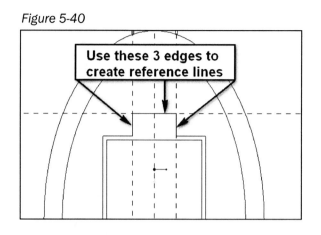

Draw a rectangle spanning these reference lines that is **.15** inch wide (see 5-41).

Figure 5-41

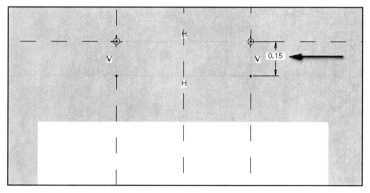

Complete the sketch.

Negatively extrude the sketch through the part.

The result is shown in Figure 5-42.

Save the part.

The final part is finished! On to the assembly!

Figure 5-42

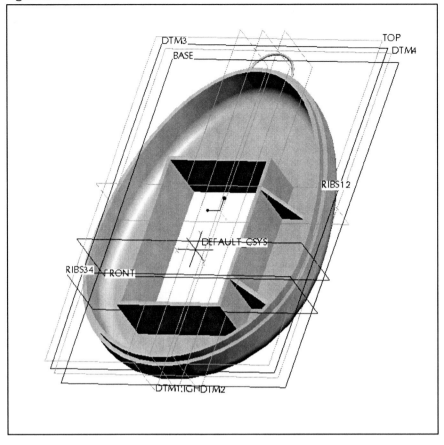

Part 6—Assembly

An assembly file combines multiple parts in the same 3-D environment. Each part is constrained to a specific location in reference to the central ASM datum planes as well as the other parts in the assembly. Although assemblies seem to be quite different than parts, they follow the same format. Each part can be thought of as a series of unique features. Each assembly is a series of constraints. The number of references needed to constrain a part and define its location completely depends on the type of reference being used.

As parts are brought into the assembly, they form a model tree like the successive features did in part modeling. Just as with features, new parts can sometimes be dependent on parts that were brought in earlier. Each time a new part is introduced, the user has the option to base constraints off of the default ASM datum planes or the other parts in the assembly. Thus, changing the position of an earlier part can cause the location of a later part to change with it.

Since every part exists in the assembly in reference to the other parts, inconsistency in units will quickly become apparent. If you add a new part to the assembly and are unable to find it, use the **Refit to Screen** icon. If the part is far too small or large, go back to the original part file and change the units.

Set the working directory. It is important that all of the parts accompany the assembly file in the same folder.

Create a new assembly. Select **File** > **New**. Choose **Assembly** as the **Type** and **Design** as the **Sub-Type** (see Figure 6-1). Name the new file **Assembly** and make sure that the **Use default template** option is active.

Figure 6-1

Notice that the base datum planes and central coordinates systems are preceded by **ASM**. As new parts are brought into the assembly, they bring their own sets of base datum planes. The **ASM** prefix mitigates confusion between the part's base references and those belonging to the assembly file.

Make sure that all of the visibility options in the top toolbar are **deactived** (not highlighted, see Figure 6-2) before continuing. The datum features are useful when in assembly mode, but will be activated as needed to minimize confusion.

Select the **Assemble** icon (see Figure 6-3) in the vertical toolbar on the right.

Figure 6-3

Figure 6-2

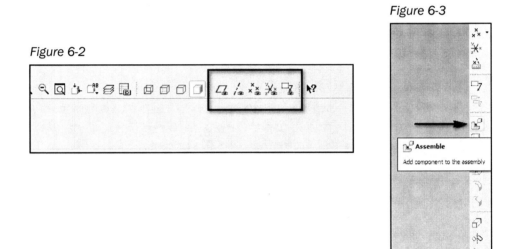

Locate the file **base.prt** (see Figure 6-4). **Double-click** it or select it and choose **Open**.

Figure 6-4

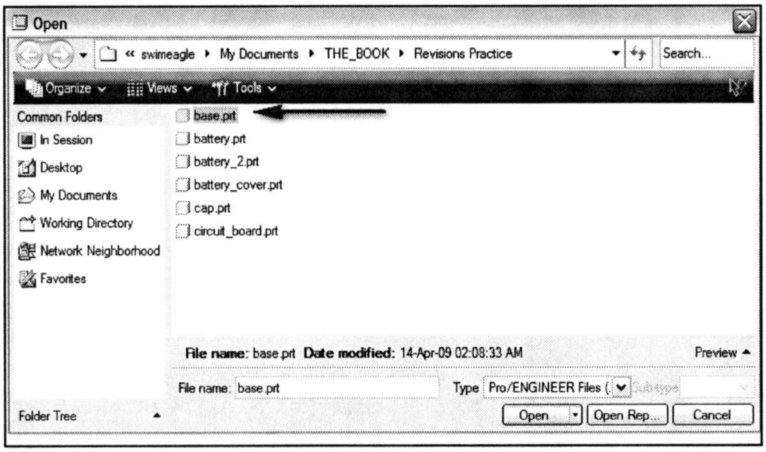

The **Base** will be constrained to the **ASM** datum planes using the **Default**

option (see Figure 6-5) to establish a basis for the other parts coming into the assembly.

Select **Default** from the bottom of the right drop-down list at the top of the screen. The part **Status** is automatically **fully constrained** (see Figure 6-6).

Figure 6-5

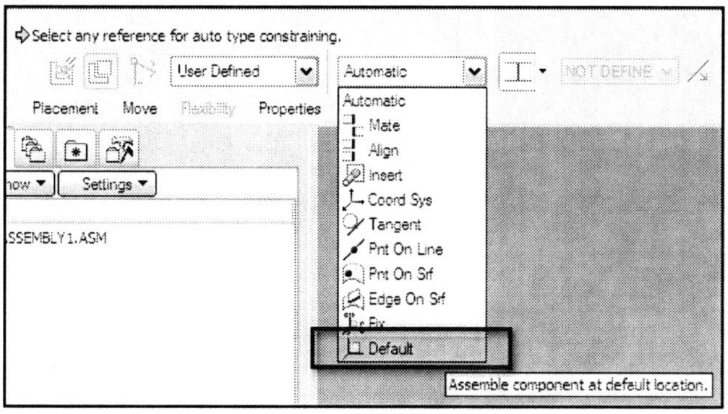

Figure 6-6

Middle-click or select the **Green Check** to complete the constraint.

The base of the assembly is set. Now additional parts can be brought into the environment and constrained off of the base part in addition to the **ASM** planes.

Select the **Assemble** icon again. Choose **circuit_board.prt** to bring the object into the assembly environment.

Choose **Surface** in the **Selection Filter** (see Figure 6-7) in the lower left as was done when creating parts.

Figure 6-7

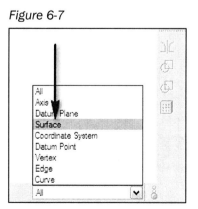

Choose **Mate** from the **Constraint Type** drop-down menu at the top left (see Figure 6-8).

Figure 6-8

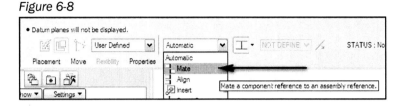

Select the large flat surface of the **Circuit Board**. Then select the thin rectangular top surface of the battery casing in the **Base** (see Figure 6-9).

Figure 6-9

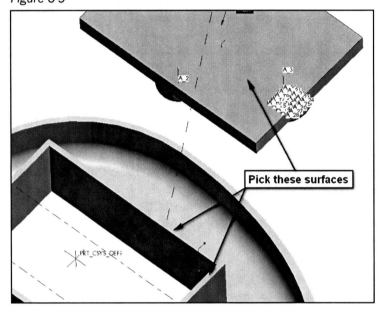

Pick these surfaces

The **Circuit Board** may have flipped over depending on its orientation coming into the environment.

Select the **Placement** tab in the upper left indicated in Figure 6-10. Choose **New Constraint** (see Figure 6-11).

Figure 6-10

Figure 6-11

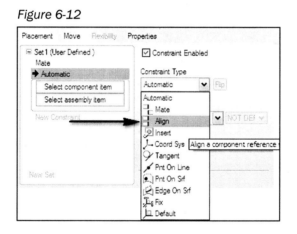

Select **Align** (Figure 6-12) from the new **Constraint Type** drop-down list. Also, make sure that **Coincident** is chosen from the **Offset** drop-down menu located beneath **Constraint Type**.

Figure 6-12

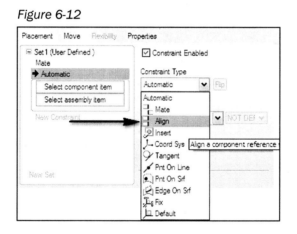

Set the **Selection Filter** to **Surface**. Choose the surfaces shown in Figure 6-13.

Figure 6-13

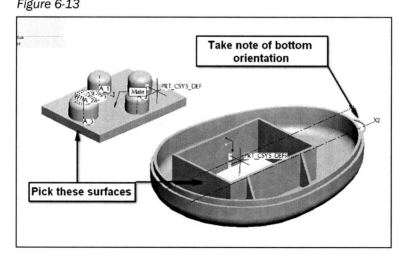

Again, choose **New Constraint** and set the **Selection Filter** to **Surface**. Set the **Constraint Type** to **Align**. Pick the surfaces shown in Figure 6-14. Again, make sure that **Coincident** is set under **Offset**.

Figure 6-14

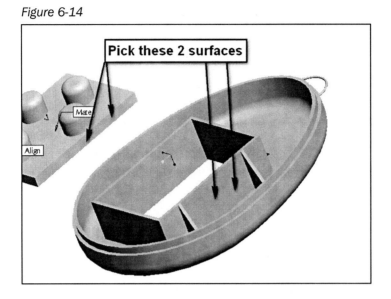

The **Circuit Board** is now constrained in all 3 dimensions **(fully constrained)**. It cannot deviate from its current position without violating one of the constraints that was previously set forth. When exclusively using mating and alignment constraints, 3 constraints will be necessary.

Check the final product on your screen against the one shown in Figure 6-15.

Figure 6-15

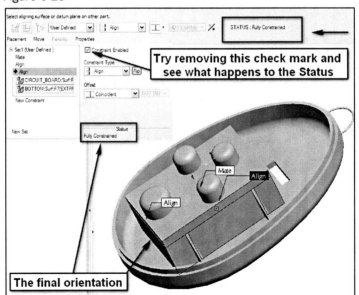

Notice the **Constraint Enabled** toggle within the **Placement** box. Uncheck the box. Notice that the **Status** changes from **fully constrained** to **partially constrained**. Sometimes constraints will conflict with one another and it is useful to disable one constraint to see what happens to the overall orientation of the part. Recheck the box.

Middle-click or press the **Green Check** to complete the assembly of the **circuit board**.

Also note that the **Mate** and **Align** constraints are very similar. The options differ by a rotation of 180 degrees. For example, if you were to select the two palms of your hands with **Mate** as the **Constraint Type**, your hands would go together as though you were clapping. With **Align** chosen as the **Constraint Type**, the hands would level with both palms facing the same direction (like when a referee calls pass interference in football).

Bring the **Battery Cover** into the environment just as with the previous two parts. Be sure to continually set the **Selection Filter** to **Surface** each time a new constraint is applied to the **Battery Cover**. Select the surfaces shown in Figure 6-16.

Figure 6-16

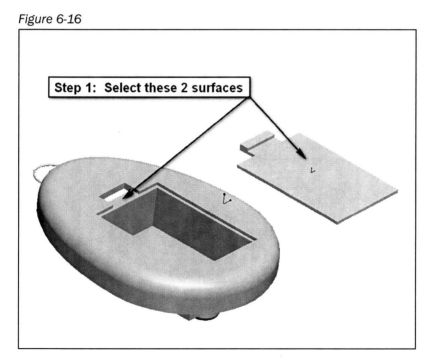

Enter the **Placement** tab. Notice how the constraint shown here is set to **Align**. The two surfaces need to be **Mated**. Also, make sure that the **Offset** is set to **Coincident** (see Figure 6-17).

Figure 6-17

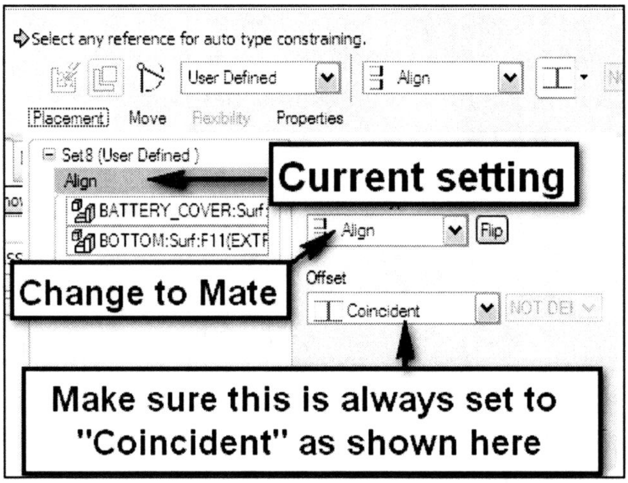

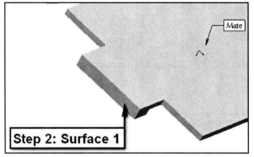

Choose **New Constraint** and select the 2 surfaces shown in Figures 6-18 and 6-19.

Figure 6-18

Figure 6-19

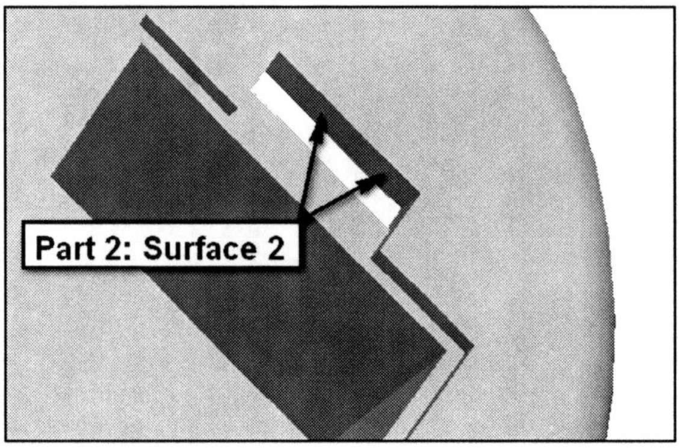

Set the **Constraint Type** to **Mate**.

Again, choose **New Constraint** to establish the third and final constraint.

Select the 2 surfaces shown in Figures 6-20 and 6-21.

Figure 6-20

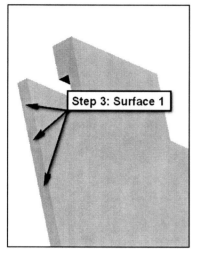

Figure 6-21

Make sure that the **Offset** is still set to **Coincident** (see Figure 6-22) and the **Constraint Type** is set to **Mate**.

Figure 6-22

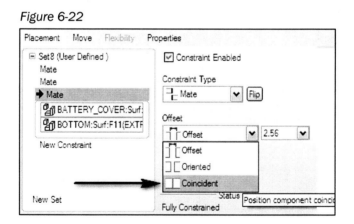

Complete the assembly of the **Battery Cover** (see Figure 6-23).

Figure 6-23

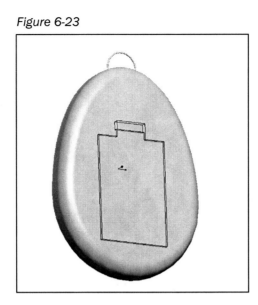

Notice that the **Constraint Type** is set to **Automatic** unless you designate it otherwise. For each of the selections just made, **PE4** would automatically choose either a **Mate** or **Align** constraint because surfaces were being used.

The **Model Tree** now contains the three parts that have been added to the assembly. Suppose that the constraints on **Battery Cover** were incorrect. A part's assembly can be edited as easily as its feature.

Right-click BATTERY_COVER in the **Model Tree**. Choose **Edit Definition**.

Open the **Placement** tab and **right-click** the third **Mate** constraint (see Figure 6-24). Select **Disable**.

Figure 6-24

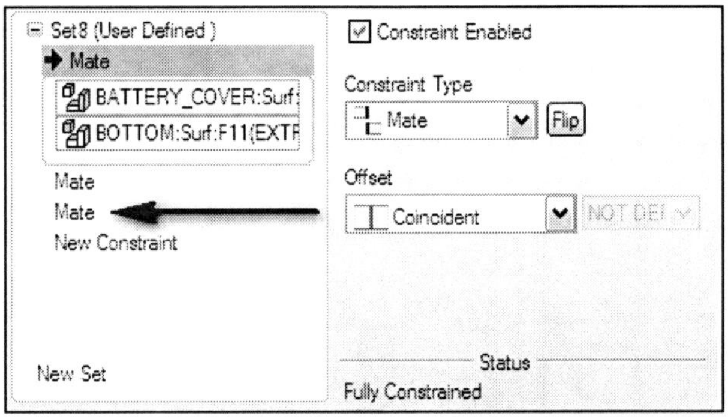

Next, the **Battery Cover** will be moved out of position. The move function in PE4 will only allow the part to slide along a plane since it is still constrained in two other dimensions.

Choose the **Move** tab shown in Figure 6-25. Click the **Battery Cover** in the assembly environment and move the mouse to the right until the **Battery Cover** is no longer in contact with the **Base**. Click again to end the movement.

Figure 6-25

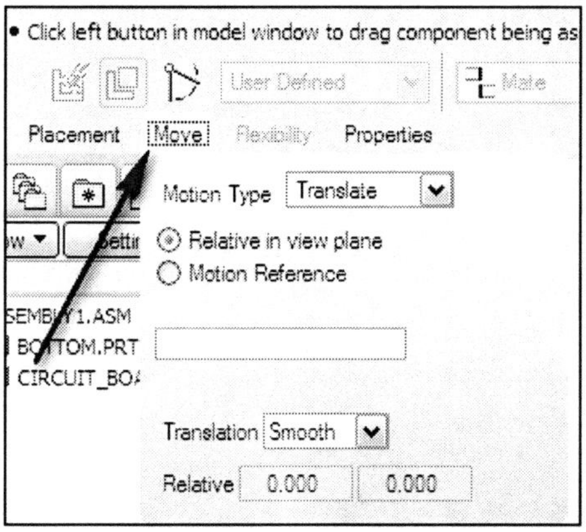

Reselect the **Placement** tab. If an incorrect surface is ever selected when in assembly mode there are two courses of action. The entire constraint can be deleted (using a right-click), or the surface can be removed and reselected. Note that there are two combinations of surfaces that could have been used to achieve this final **Mate** constraint. It is currently left surface mated to left surface, but it just as easily could have been right surface mated to right surface. Here, the left surfaces are removed and replaced with the right surfaces as practice.

Right-click the item labeled **BATTERY_COVER: Surf** under the third **Mate** constraint. Choose **Remove** (see Figure 6-26).

Figure 6-26

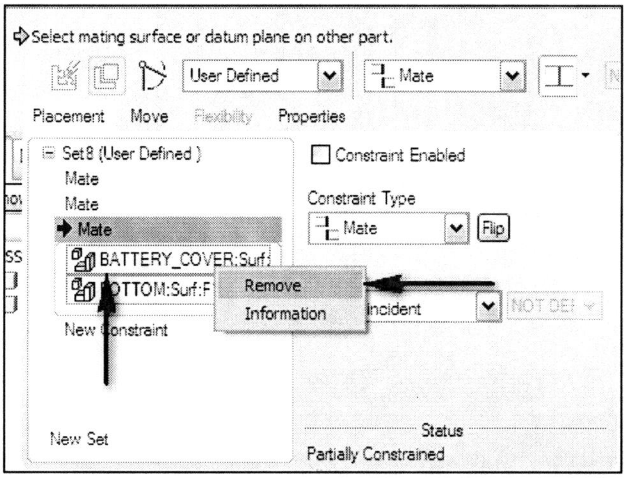

Select the right surface of the **Battery Cover** (see Figure 6-27).

Figure 6-27

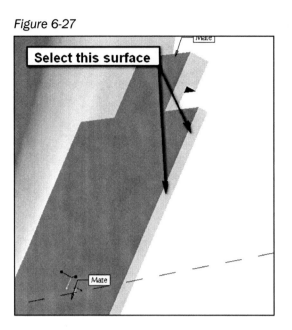

Similarly, **Remove** the second item under the third **Mate** constraint. Select the surface right edge of the **Base** (see Figure 6-28).

Figure 6-28

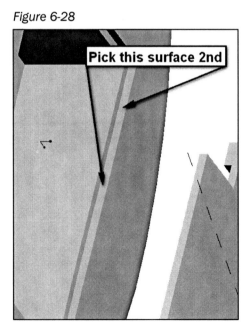

Check the **Constraint Enabled** box under the **Placement** tab. **Complete** the reassembly.

Next, the **Batteries** will be inserted. The surfaces and axes needed to constrain the **Batteries** are now hidden by the **Battery Cover** and the **Circuit Board**. Fortunately, **PE4** has accounted for this nasty situation.

In the **Model Tree**, **right-click** the **BATTERY_COVER** and select **Hide** (see Figure 6-29).

Figure 6-29

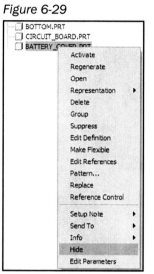

Activate the **Axis Display** (see Figure 6-30). The datum axes that were created in Part 5 with the **Base** will now become useful.

Figure 6-30

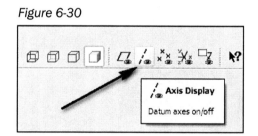

Both **Battery** and **Battery_2** are the same. Either of the two parts can be used here.

Bring a **Battery** into the assembly.

Leave the **Constraint Type** set to **Automatic**. Choose either of the central axes in the **Battery (A_1 or A_2)**. Choose the **X_1** axis in **Bottom** (see Figure 6-31).

Figure 6-31

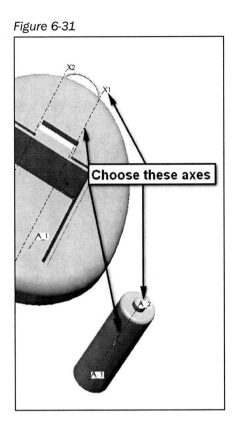

Mate the top surface of the **Battery** to the vertical wall in the **Base** closest to the key ring (see Figures 6-32, 6-33).

Figure 6-32   *Figure 6-33*

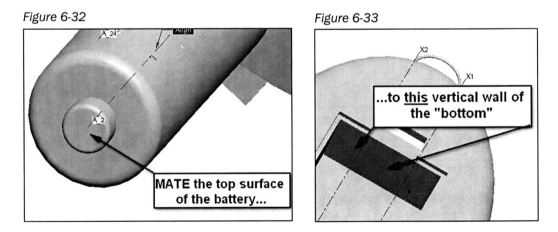

Again, be sure the **Offset** is set to **Coincident**. The desired result is shown in Figure 6-34.

Figure 6-34

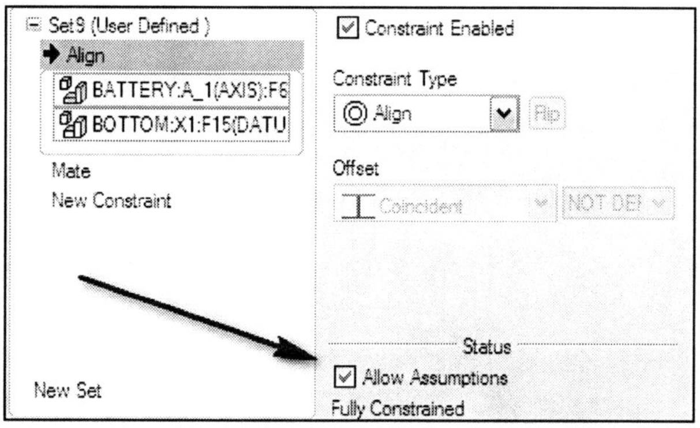

Set 9 (User Defined)
→ Align
BATTERY:A_1(AXIS):F8
BOTTOM:X1:F15(DATU
Mate
New Constraint

New Set

☑ Constraint Enabled

Constraint Type
◎ Align ▼ Flip

Offset
⊥ Coincident ▼ NOT DEF ▼

Status
☑ Allow Assumptions
Fully Constrained

Notice how the **Battery** only required two constraints to be **fully constrained**. When the two axes were aligned, the **Battery** could only move in one dimension, back and forth along the now common axis. Once the surfaces were mated, the **Battery** had zero degrees of freedom.

However, the battery could still rotate about the axis. **PE4** automatically assumes a rotational orientation that can be overridden in the **Placement** tab. Notice the **Allow Assumptions** toggle under the **Align** constraint. If the box is unchecked, the **Battery** will shift **Status** to **partially constrained** since there is a new degree of freedom (rotation) present in the system. Leave the box checked as the rotation does not matter since the battery is symmetrical about the full 360 degrees.

Complete the assembly of the **battery**.

Assemble a second **battery** so that the final result looks like Figure 6-35.

Figure 6-35

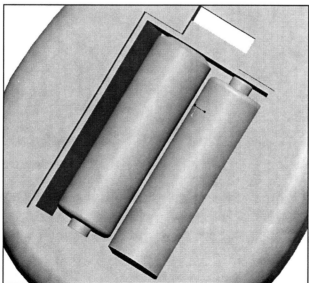

In the **Model Tree**, **right-click BATTERY_COVER** and choose **Unhide**.

Finally, import the **Cap** into the assembly. Turn the **visibility ON for datum planes** and **OFF for datum axes** (see Figure 6-36).

Figure 6-36

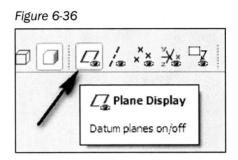

Move the **Cap** slightly away from the **Bottom** so that the datum planes are less jumbled.

Mate the lip of the **Cap** (see Figure 6-37) with the ridge of the **Bottom** (see Figure 6-38).

Figure 6-37

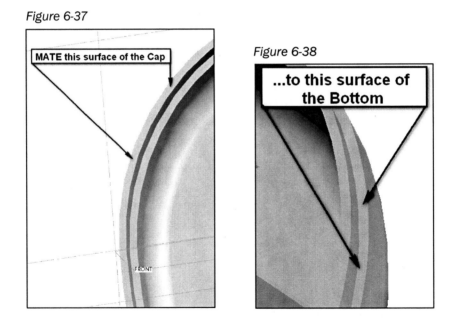

Figure 6-38

Align the **FRONT** datum plane of the **Cap** with the **ASM FRONT** (see Figures 6-39, 6-40). The **FRONT** of the **Base** can also be used since it is constrained to be co-planar with **ASM FRONT**. Treat the datum planes as though they were surfaces.

Figure 6-39

Figure 6-40

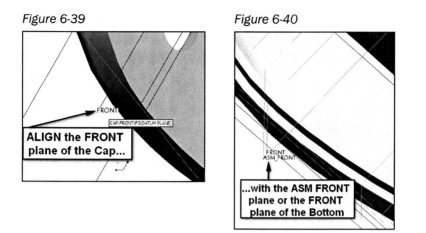

Mate the **RIGHT** datum plane of the **Cap** to the **ASM RIGHT** datum plane (see Figure 6-41).

Figure 6-41

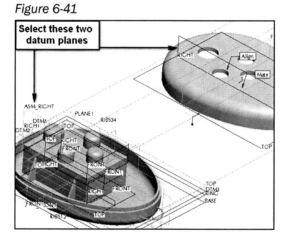

Complete the assembly of the **Cap**.

Turn off the datum plane visibility. Observe the finished product thoroughly. Do you notice any defects? There are gaps between the buttons of the **Circuit Board** and the **Cap** (see Figure 6-42)!

Figure 6-42

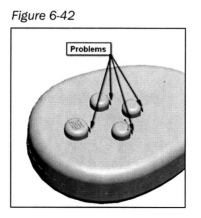

Assemblies can apply live updates to parts. In other words, if a part in the assembly is opened separately and edited, the changes will instantly transfer to the part appearing in the assembly. The **Cap** must be edited so that holes will yield a tight fit.

Open **cap.prt** as a separate part file. Locate and **right-click Sketch 2** in the **Model Tree**. Select **Edit Definition**.

The corrected sketch dimensions are shown in Figure 6-43.

Figure 6-43

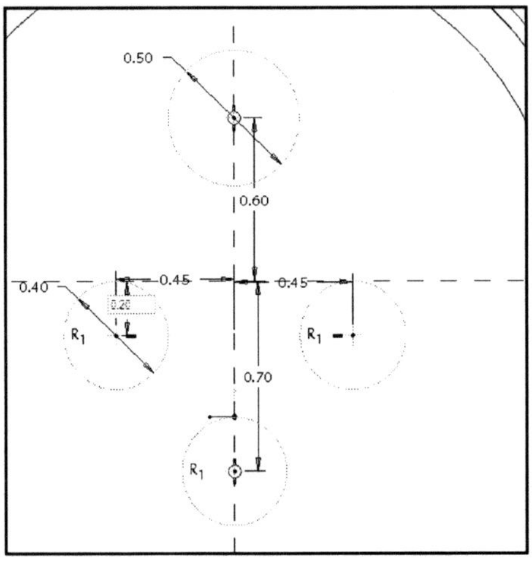

The following dimension **changes** need to be made:

OLD		NEW
.25	→	.2
.55	→	.6
.75	→	.7

When the dimensions have been edited, **complete** the sketch. The part regenerates. **Save** the part.

Activate the **Assembly** window (**Control + A**).

Notice how the assembly automatically updates and the part is corrected!

Finally, color will be added to each of the parts to make them more distinct in the assembly.

Activate the **Cap** window.

Select **View** > **Color and Appearance** from the top menu (see Figure 6-44).

Click on the **Cap**. Select the **+ (Add new Appearance)** shown in Figure 6-45 followed by the **Color Box.**

Figure 6-44

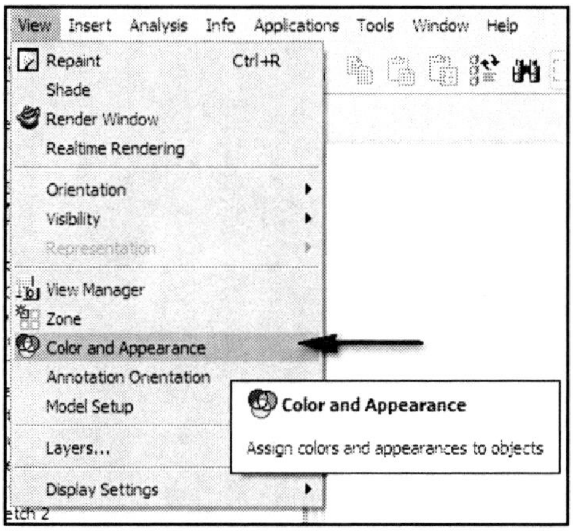

Figure 6-45

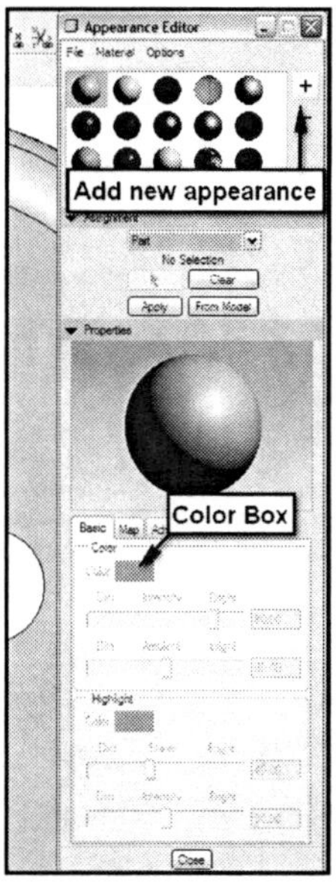

Click and drag the **RGB** bars in the **Color Editor** to get a new color. Create something different than the current gray-blue. When finished, select **Close** in the **Color Editor** and choose **Apply** under **Assignment** in the **Appearance Editor** (see Figure 6-46).

Figure 6-46

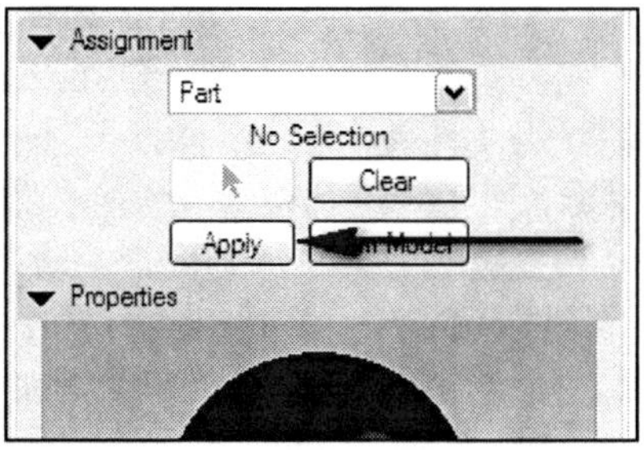

Close the **Appearance Editor**.

Again, **Activate** the **Assembly**. The **Cap** appears in its new color.

Change the color of the rest of the parts to something unique so that each part is distinct in the assembly.

Congratulations on completing the remote control assembly! Your **PE4** skills have come so far. However, there is still much more to learn. Hone your skills by trying to model the things around you, perhaps a pencil or a calculator. Also, experiment with other tools that were not discussed in this tutorial. Remember, the **message area** can be a helpful guide.

About the Authors

Avery J. Scott is pursuing a Bachelor's degree in Mechanical Engineering as a member of the class of 2010 at the University of Notre Dame. He has had experience working with various CAD software packages while with Rolls-Royce North America in the summer of 2008. His background in high school education stems from his experience as a mentor for FIRST Robotics (2007-2009) and a guest lecturer during CAD instruction (2008) at St. Edward High School in Lakewood, OH. He currently serves as a volunteer assistant to the engineering program at Riley High School in South Bend, IN.

Richard B. Strebinger is an instructor in the Department of Aerospace and Mechanical Engineering at the University of Notre Dame. He teaches a junior level course in Computer Aided Design and Manufacturing using Pro/ENGINEER software. Along with providing classroom and laboratory instruction for students, Strebinger assists faculty and graduate students in the design and fabrication of mechanical devices used in research and teaching. He also manages the Rapid Prototyping Laboratory and CAD/CAM facilities used in the AME Department. He received a Bachelor's degree from Tri-State University in 1981 and a Master's degree from Rensselaer Polytechnic Institute in 1983.

Printed in the United States
217034BV00002B/1/P

9 780978 879396